CONTENTS (continued)

Title	Section	Page
Light, Lighter, Lightest Heavy, Heavier, Heaviest	Weight	T-2
Light, Lighter, Lightest, Heavy, Heavier, Heaviest	Weight	2
Weigh with Nonstandard Units	Weight	T-3
Weigh with Bears	Weight	3
More or Less than 1 Pound	Weight	T-4
More or Less than 1 Pound	Weight	4
About 1 Pound	Weight	T-5
About 1 Pound	Weight	5
More or Less than 1 Kilogram	Weight	T-6
Kilogram Crush	Weight	6
Read and Write About Weight	Weight	T-7
Light and Heavy	Weight	7
Unit Answer Key	Weight	T-10
Temperature Intro	Temperature	T-1A
Temperature Unit Manips List	Temperature	T-1B
Get to Know Temperature	Temperature	T-1
Temperature or Not?	Temperature	1
Use Temperature Words	Temperature	T-2
Hot, Cold, or in Between?	Temperature	2
Compare Temperature	Temperature	T-3
Where Is It Hot?	Temperature	3
Order Temperatures	Temperature	T-4
Coldest to Hottest	Temperature	4
Read a Thermometer	Temperature	T-5
Draw Me	Temperature	5
Estimate Temperature	Temperature	T-6
About How Hot?	Temperature	6
Measure in Degrees Fahrenheit	Temperature	T-7
Explore Outdoors	Temperature	7
A Week of Weather	Temperature	T-8
A Week of Weather	Temperature	8
Unit Answer Key	Temperature	T-11
Assessment	Assessment	T-1A
Time	Assessment	T-1
Length	Assessment	T-2
Capacity	Assessment	T-3
Weight	Assessment	T-4
Temperature	Assessment	T-5
Comprehensive Assessment	Assessment	T-6
Reproducible Materials		
Measurement Masters	BLM	1
KWL Chart	BLM	7

Title	Section	Page
Overhead Masters	BLM	8
MeasureWorks Games	BLM	13

MeasureWorks™ • Grade 1

MeasureWorks™ Program Components

This grade 1 Teacher's Resource Guide provides the materials you need for effective teaching and learning.

- **Teacher's Guide**
    Daily Lesson Plans for
        Time
        Length
        Capacity
        Weight
        Temperature
    Unit Projects
    Assessment and Standardized Test Preparation
    Reproducible Measurement Masters

- **Student Activity Pages**
    reproducible OR
    consumable Student Activity Workbooks

- **MeasureWorks™ Game Boards**

- **MeasureWorks™ Tool Kit**

Program Consultant: Dr. Carol Thornton, Distinguished Professor, Mathematics Department, Illinois State University, Normal, Illinois
Product Development Manager: Mary Watanabe
Project Coordinators: Kevin Anderson, Jennifer Martin
Lead Editor: Bill Gasper
Editorial Team: Brion McGinn, Nancy Sheldon
Editorial, Design, and Production Team: Christine Losq Education Associates, Palo Alto, California

MeasureWorks™ Teacher's Resource Guide, Grade 1
ISBN 0-7406-0370-1
ETA 40981TG

ETA/Cuisenaire • Vernon Hills, IL 60061-1862
800-445-5985 • www.etacuisenaire.com

© 2003 by ETA/Cuisenaire®
All rights reserved.

No part of this publication may be reproduced, stored in a retrieval system, or transmitted, in any form or by any means, electronic, mechanical, photocopying, recording, or otherwise, without the prior written permission of the publisher. Permission is granted for limited reproduction of those pages from this book with a copyright line in the inner margin for classroom use and not for resale.

Printed in China.

03 04 05 06 07 08 09 10 11 12   10 9 8 7 6 5 4 3 2 1

CONTENTS

Title	Section	Page
Introduction to MeasureWorks		iv
Scope and Sequence		xiv
Time Intro	Time •	T-1A
Time Unit Manips List	Time •	T-1B
Review Time Vocabulary	Time •	T-1
Clock and Calendar	Time •	1
Days-of-the-Week Memory	Time •	T-2
Remember the Days!	Time •	2
Name and Order Months of the Year	Time •	T-3
My Seasons Book	Time •	3
Daily Calendar Activity	Time •	T-4
My Calendar	Time •	4
Estimate Time to the Hour	Time •	T-5
It's About Time	Time •	5
Tell Time Before and After the Hour	Time •	T-6
Color, Show, Write	Time •	6
Explore the Minute Hand	Time •	T-7
Who Has Playtime?	Time •	7
Tell Time to the Half-Hour	Time •	T-8
Hour and Half-Hour	Time •	8
Read Digital and Analog Clocks	Time •	T-9
Match the Clocks	Time •	9
Recognize Time to the Half-Hour	Time •	T-10
Time Match-Up	Time •	10
Estimate Elapsed Time	Time •	T-11
Just a Minute	Time •	11
Use a Schedule	Time •	T-12
Camp Sunny	Time •	12
Make a Schedule	Time •	T-13
My Schedule	Time •	13
Am I Late?	Time •	T-14
Am I Late?	Time •	14
Unit Answer Key	Time •	T-17
Length Intro	Length •	T-1A
Length Unit Manips List	Length •	T-1B
Get to Know Length	Length •	T-1
PopCube Trains	Length •	1
Understand Long, Longer, Longest	Length •	T-2
Long and Tall	Length •	2
Understand Short, Shorter, Shortest	Length •	T-3
Short, Shorter, and Shortest	Length •	3
Compare Length	Length •	T-4
Compare Length	Length •	4
Measure with Link 'N' Learn Links	Length •	T-5
Clay Snakes	Length •	5
Explore Units of Measure	Length •	T-6

Title	Section	Page
Ways to Measure	Length •	6
Measure with Steps	Length •	T-7
Step by Step	Length •	7
Estimate with Nonstandard Units	Length •	T-8
Grab Bag	Length •	8
Estimate and Check	Length •	T-9
Estimate and Check	Length •	9
Measure with Inchworms	Length •	T-10
Inchworms	Length •	10
Measure with Rulers	Length •	T-11
Measure with Rulers	Length •	11
Measure in Centimeters	Length •	T-12
Centibug Straws	Length •	12
Estimate in Centimeters	Length •	T-13
Sliding Links	Length •	13
Measurement Man	Length •	T-14
Measurement Man	Length •	14
Unit Answer Key	Length •	T-17
Capacity Intro	Capacity •	T-1A
Capacity Unit Manips List	Capacity •	T-1B
Get to Know Capacity	Capacity •	T-1
How Many to Fill It?	Capacity •	1
Compare Capacity Using Less Than or More Than	Capacity •	T-2
More or Less?	Capacity •	2
Order Containers Using More, Most, Less, and Least	Capacity •	T-3
Less, More, the Most	Capacity •	3
Estimate Capacity	Capacity •	T-4
How Much?	Capacity •	4
Explore Standard Units of Capacity	Capacity •	T-5
Exact or About?	Capacity •	5
Explore Math Tools: One Cup	Capacity •	T-6
Color a Cup	Capacity •	6
Estimate and Measure One Cup	Capacity •	T-7
About a Cup?	Capacity •	7
Explore Math Tools: One Liter	Capacity •	T-8
Follow the Liter	Capacity •	8
Goldilocks Comes Back	Capacity •	T-9
Goldilocks Comes Back	Capacity •	9
Unit Answer Key	Capacity •	T-11
Weight Intro	Weight •	T-1A
Weight Unit Manips List	Weight •	T-1B
Lighter, Heavier, the Same	Weight •	T-1
Lighter or Heavier	Weight •	1

ii

MeasureWorks™ • Grade 1

MeasureWorks™ Grade 1 Tool Kit

includes the following manipulatives:

	Quantity	Time	Length	Capacity	Weight	Temperature
Big Time™ Learning Clock®	1	●				
Geared Student Clock	24	●				
Link 'N' Learn® Links	500		●			
Inchworms™	4 sets		●			
Inchworms™/Centibugs™ Ruler	24		●			
Folding Meter Stick	1		●			
Dual Measuring Cups	6 sets			●		
PopCubes®	400		●	●		●
Baby Bear™ Counters	200			●	●	
Precision School Balance	4				●	
Standard Mass Set	4				●	
Kilogram Mass	4				●	
Student Thermometer	12					●
Classroom Thermometer	1					●

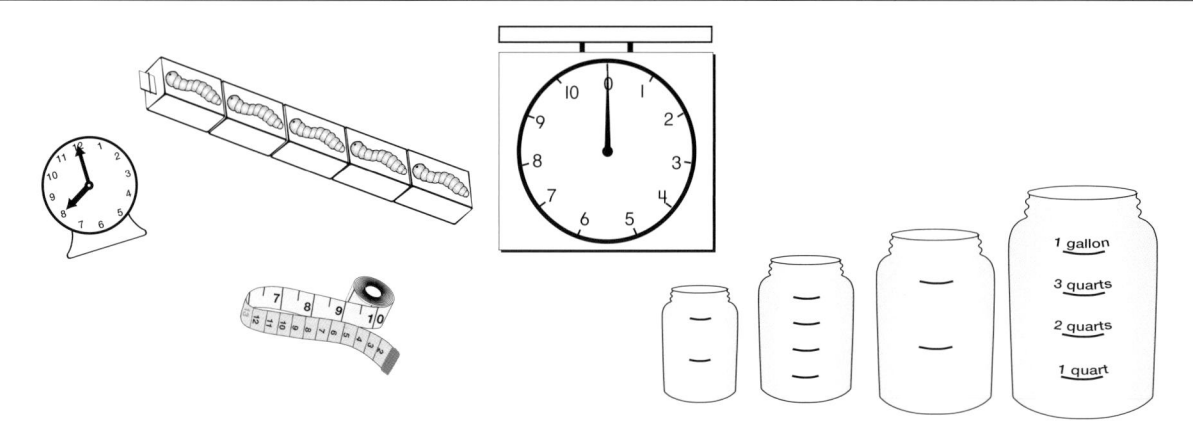

MeasureWorks™ • Grade 1

v

MeasureWorks

Grade 1
Teacher's Resource Guide

The following measurement topics are studied in MeasureWorks:

• Time • Length • Capacity

• Weight • Temperature

Use MeasureWorks as a stand-alone program

OR

pick and choose MeasureWorks activities for learning centers to target your students' needs and supplement your basal text.

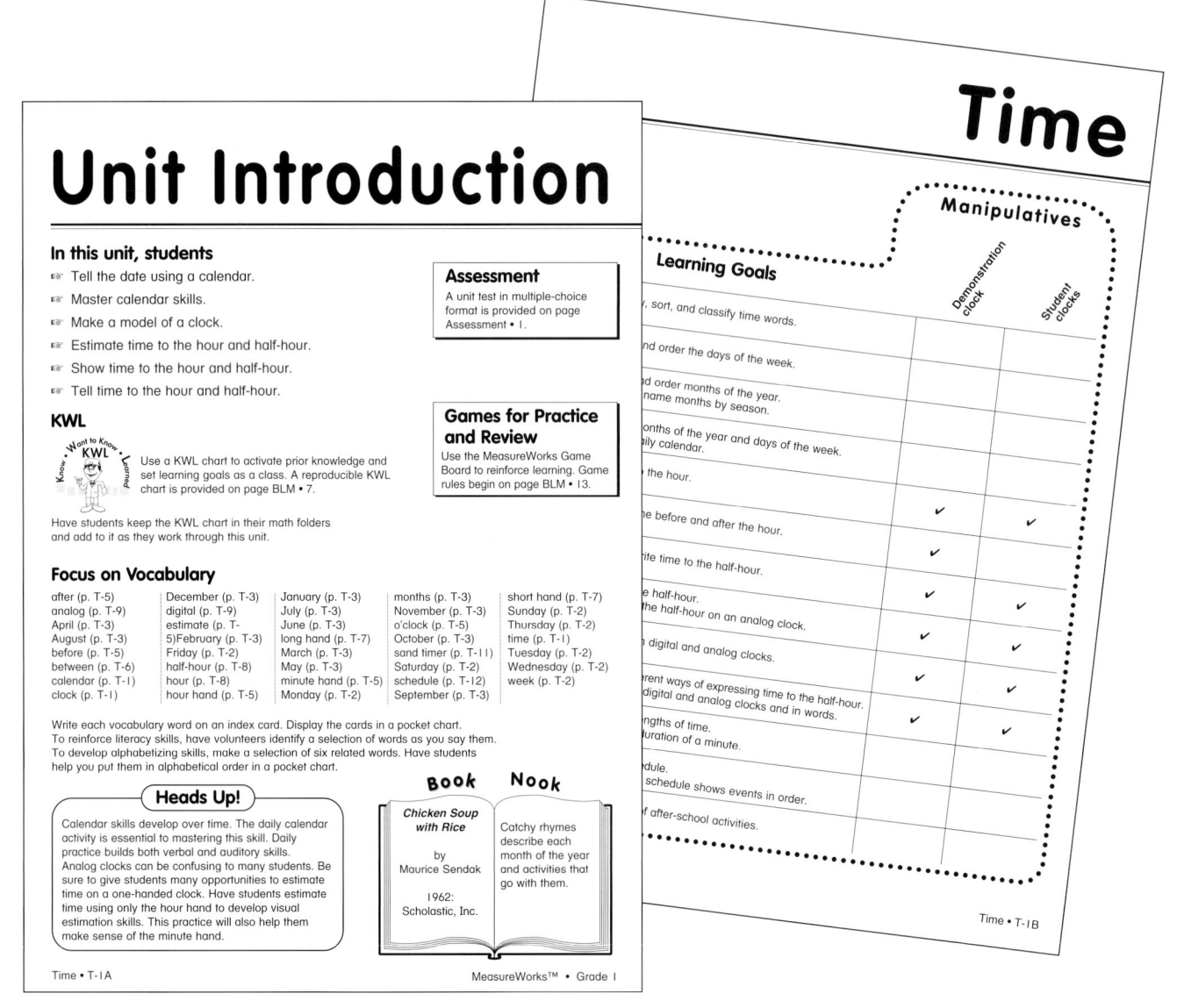

Unit Introduction

In this unit, students
- Tell the date using a calendar.
- Master calendar skills.
- Make a model of a clock.
- Estimate time to the hour and half-hour.
- Show time to the hour and half-hour.
- Tell time to the hour and half-hour.

Assessment
A unit test in multiple-choice format is provided on page Assessment • 1.

KWL

Use a KWL chart to activate prior knowledge and set learning goals as a class. A reproducible KWL chart is provided on page BLM • 7.

Have students keep the KWL chart in their math folders and add to it as they work through this unit.

Games for Practice and Review
Use the MeasureWorks Game Board to reinforce learning. Game rules begin on page BLM • 13.

Focus on Vocabulary

after (p. T-5)	December (p. T-3)	January (p. T-3)	months (p. T-3)	short hand (p. T-7)
analog (p. T-9)	digital (p. T-9)	July (p. T-3)	November (p. T-3)	Sunday (p. T-2)
April (p. T-3)	estimate (p. T-5)	June (p. T-3)	o'clock (p. T-5)	Thursday (p. T-2)
August (p. T-3)	February (p. T-3)	long hand (p. T-7)	October (p. T-3)	time (p. T-1)
before (p. T-5)	Friday (p. T-2)	March (p. T-3)	sand timer (p. T-11)	Tuesday (p. T-2)
between (p. T-6)	half-hour (p. T-8)	May (p. T-3)	Saturday (p. T-2)	Wednesday (p. T-2)
calendar (p. T-1)	hour (p. T-8)	minute hand (p. T-5)	schedule (p. T-12)	week (p. T-2)
clock (p. T-1)	hour hand (p. T-5)	Monday (p. T-2)	September (p. T-3)	

Write each vocabulary word on an index card. Display the cards in a pocket chart. To reinforce literacy skills, have volunteers identify a selection of words as you say them. To develop alphabetizing skills, make a selection of six related words. Have students help you put them in alphabetical order in a pocket chart.

Heads Up!
Calendar skills develop over time. The daily calendar activity is essential to mastering this skill. Daily practice builds both verbal and auditory skills. Analog clocks can be confusing to many students. Be sure to give students many opportunities to estimate time on a one-handed clock. Have students estimate time using only the hour hand to develop visual estimation skills. This practice will also help them make sense of the minute hand.

Book Nook
Chicken Soup with Rice
by Maurice Sendak
1962: Scholastic, Inc.

Catchy rhymes describe each month of the year and activities that go with them.

Time • T-1A

MeasureWorks™ • Grade 1

Time

Manipulatives: Demonstration clock, Student clocks

Learning Goals
- ...y, sort, and classify time words.
- ...nd order the days of the week.
- ...d order months of the year.
- ...name months by season.
- ...onths of the year and days of the week.
- ...ily calendar.
- ...the hour.
- ...e before and after the hour.
- ...ite time to the half-hour.
- ... half-hour.
- ...the half-hour on an analog clock.
- ...digital and analog clocks.
- ...rent ways of expressing time to the half-hour.
- ...digital and analog clocks and in words.
- ...ngths of time.
- ...uration of a minute.
- ...dule.
- ...schedule shows events in order.
- ...f after-school activities.

Time • T-1B

MeasureWorks Lesson Plans at a Glance

The easy-to-use lesson plans in MeasureWorks were designed to minimize preparation time by providing helpful management tips, specific questions for students, and cross-curricular extensions.

Plan the time you need for each part of the lesson.

Plan your lesson set-up.

Read Digital and Analog Clocks

Planning Your Time
- Intro & Demo: 15 min
- Activity: 15 min
- Sum It Up: 10 min

Objective
Match times on digital and analog clocks.

Materials
- Student clocks
- Demonstration clock

Grouping
Whole class, then pairs

Open It Up
Show 4:00 on the demonstration clock.

Ask: Where is the hour hand? [on the 4]
Where is the minute hand? [Sample: on the 12]

Say: When the minute hand is pointing straight up to the 12, it shows zero minutes after the hour. We say the time is 4 o'clock.

Write *4 o'clock* and *4:00* on the board.

Say: Our clock shows 4 o'clock. We can write it as 4:00, which means zero minutes after the hour. Repeat with other examples, such as 6:00 and 9:00.

Say: A clock that has hands and numbers going around it is called an **analog** clock.

Demonstrate & Discuss
Draw a digital clock face on the board. Tell students that a clock that shows time this way is called a digital clock.

Show 8:00 on the demonstration clock. Have a volunteer tell what time is shown. [8:00] Write 8:00 on the digital clock face.

Ask: What time did I write on the board? [8 o'clock] How many minutes past the hour does my clock show? [zero] Where is the minute hand? [on the 12] Now turn the minute hand to show 8:30. Write 8:30 on the digital clock face.

Ask: What time does my clock show now? [30 minutes past 8 o'clock] What time did I write on the board? [8:30] How many minutes past the hour is 8:30? [30 minutes]

Student Activity
Prepare ahead: Each student will need a student clock.

Read the directions on the student page aloud to students. Students read time to the hour and half-hour in digital form. They write how many minutes past the hour each time shows. They they model the time on student clocks and draw the hands of an analog clock to show the time.

Informal Assessment
As students work, help them use their listening skills and time vocabulary.

Ask: Where is the minute hand when the time is 7 o'clock? [Sample: on the 12] /DESCRIBE/

Where is the hour hand? [on the 7] /DESCRIBE/

Sum It Up
Say: Today we learned that there are two ways to tell time. A digital clock tells time in digits. An analog clock tells time with hands on a clock face.

Ask: Where is the minute hand when the time is 7:30? [on the 6] /DESCRIBE/

Where is the hour hand when the time is 7:30? [between the 7 and the 8] /DESCRIBE/

Literature Connection
The Carrot Seed by Ruth Krauss (Harper and Row, 1945)

Engage students to activate prior knowledge.

Model the activity before students explore and record.

Students do, show, and tell.

Help students actively use vocabulary.

Summarize learning.

Connect across the curriculum.

Reproducible student pages and Measurement Masters help you evaluate student learning.

Name _____

Match the Clocks

Try This
- Read the time on the digital clock.
- Write the time.
- Show the same time on your clock.
- Draw the time.

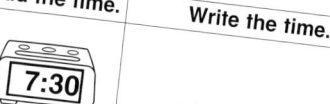

At zero minutes past the hour, the minute hand is at 12.

① Read the time. | Write the time. | Draw the time.
7:30 | ___ minutes past ___ o'clock |

②

MeasureWorks™ • Grade 1

vii

Teaching with MeasureWorks

Introduction by Carol Thornton, Ph.D.

> Measurement is one of the most widely used applications of mathematics.
>
> – *Principles and Standards for School Mathematics* (NCTM, 2000) page 44

> To enhance and refine children's knowledge of measurement:
>
> - Provide hands-on, "involving" experiences.
> - Make verbal and written language integral parts of measurement experiences.
> - Compare and estimate, then measure.
> - Have children make some of their own measurement tools.
>
> –Jean Shaw and Mary Jo Puckett Cliatt
> *New Directions for Elementary School Mathematics* (NCTM Yearbook)

How old?

How long?

How far?

How heavy?

How full?

Who hasn't had to answer questions like these both in everyday life and on standardized tests? Students need to have a working, hands-on knowledge of measurement concepts and tools to answer these questions in meaningful ways. Students must recognize the attribute to be measured, know how to measure the attribute, understand how to manipulate the numbers involved, and utilize good problem-solving skills.

In the primary- and intermediate-level measurement sections of *Principles and Standards for School Mathematics*, the National Council of Teachers of Mathematics (NCTM) emphasizes the importance of developing measurement concepts and related skills through "direct experiences" in which students are "actively involved" and draw on "familiar and accessible contexts" (NCTM, 2000).

MeasureWorks, the first curriculum of its kind, responds to NCTM's recommendations in two different ways—it can be used as a comprehensive, hands-on, measurement replacement module or as a supplement to the measurement lessons of your mathematics or science textbook. While math teachers may elect to address all or selected units, science teachers may be particularly interested in those activities that target metric measurement and its applications. Regardless of whether MeasureWorks is used by a math or science teacher, MeasureWorks teachers become part of an active team that systematically helps increase students' understandings and skills in measurement.

Low scores on national and state-level math standardized achievement tests demonstrate that students need more coherent, better-articulated measurement programs. MeasureWorks carefully structures learning consistent with NCTM's Standards, in highly engaging and motivating lessons, and correlates to the measurement sections of key state math standards.

MeasureWorks is a spiral, hands-on curriculum. Each topic and its tools are introduced at the appropriate grade and revisited in subsequent grades at more in-depth levels. Easy-to-use hands-on activities help students actively investigate a topic as they work in pairs, in small groups, or individually to measure, record, and discuss their findings. In each activity students use appropriate measurement tools, such as tape measures or balances. All the measurement tools and teaching aids needed for the activities are provided in your MeasureWorks Toolbox.

> Teachers should provide many hands-on opportunities for students to choose tools for measuring different attributes.
> – *Principles and Standards for School Mathematics* (NCTM, 2000) page 105

"Big Ideas" of Measurement

Coordinated research conducted in the last decades of the twentieth century helped identify five basic steps in students' cognitive development of measurement concepts (Copely, 2000).

- **Recognize that objects have measurable attributes** and understand what is meant by "How tall?" "How heavy?" and other questions whose answers help describe an object.

- **Make direct comparisons of two objects** using appropriate vocabulary, such as taller, heavier, and so on.

- **Determine an appropriate unit and process** for measuring an object. After students are formally introduced to more than one unit within a system, they are encouraged to choose an appropriate unit before they begin to measure. For example, in finding the length of the classroom, many second-graders recognize that it is more efficient to use yards rather than use inches. Students are also explicitly taught how to use measurement processes that result in accurate answers. For example, when measuring how many cups of water it takes to fill a 1-gallon container, students are guided to repeatedly fill the 1-cup container with water and count and record how many cups they needed to pour to fill the container.

- **Use standard units of measurement** to record measurements, explore equivalent measures, and check estimated measures.

- **Understand and use formulas** to help count units.

> Children and adults who have measurement sense have
> - knowledge of the units appropriate for a task.
> - knowledge of the measurement process.
> - the ability to decide when to measure and when to estimate.
> - knowledge of strategies for estimating length, temperature, volume, mass, and time.
>
> –Jean Shaw and Mary Jo Puckett Cliatt
> *New Directions for Elementary School Mathematics*
> (NCTM Yearbook)

MeasureWorks

> Teachers should guide students' experiences by making the resources for measuring available.
> – *Principles and Standards for School Mathematics* (NCTM, 2000) page 103

The sequence of activities within each measurement topic in MeasureWorks fully addresses the natural development of students' understanding of measurement that is outlined above. The assessments at each grade level provide teachers with guidelines for understanding and assessing students' thinking related to important measurement concepts and skills being developed.

Students are encouraged to establish benchmarks they can use to estimate a measurement of an object or distance. MeasureWorks systematically involves students in identifying benchmarks—personal or otherwise familiar ones—that relate to key, measurable attributes of an object. For example, the distance from a door handle to the bottom of a door is about 1 meter in length. Estimated measures are helpful and often used in daily life. Becoming comfortable giving approximate measures is a skill that requires on-going nurturing. Students who have had frequent opportunities to "predict" then "measure to check," become better estimators than students who have not had these direct experiences.

Theory into Practice: Measurement Experiences in MeasureWorks

Time is an abstract and complex concept. Younger students become aware of time by sequencing events, and by studying basic units of time through a variety of activities, such as discussing time intervals (e.g., "We have 30 minutes for lunch."). In the primary grades, students focus on reading analog and digital clock times. Skip counting by 5s is encouraged to help students tell time on analog clocks to the nearest 5 minutes. Visual, auditory, and kinesthetic-tactile learning strategies in interactive settings—such as those suggested in MeasureWorks—help students successfully develop needed skills for reading clock times and for determining elapsed times.

> Estimation activities are an early application of number sense.
> – *Principles and Standards for School Mathematics* (NCTM, 2000) page 106

Time also involves larger increments, such as day, week, month, year, etc. It is important that students grasp the relationships among these units and develop the ability to convert between time units. Equally important, students need frequent experiences making, reading, and interpreting time schedules, such as a typical school day or a train schedule. Finally, the concept of elapsed time should be informally introduced to students in the primary grades and made explicit to students in the intermediate grades.

MeasureWorks

Length is the measure of something from end to end, such as the width of a table. Young students need to first learn basic rudiments of measuring lengths. 1) When using non-standard units, all units must be the same size/length; and units should be aligned end to end. 2) A ruler and object to be measured must "start" at the same place. 3) If the object or distance to be measured is longer than the tool used, then a measure-and-mark process must be reiterated. Intermediate students learn that certain length measurements have specific names such as perimeter, width, height, circumference, and distance. They also explore equivalences within the same measurement system.

Capacity refers to the three-dimensional space inside a container that can be occupied by a substance such as water or sand. The capacity of a container can be measured by the number of liquid units (e.g., fluid ounces or milliliters) that will fill it. Volume, the amount of space occupied by a three-dimensional figure, can be measured by the number of cubic units needed to replicate a given figure. Although volume and capacity are closely related, the two terms are not synonymous. Even students in the early grades can be taught to distinguish between a container's capacity and the volume of a solid through a series of hands-on investigations and direct instruction.

Activities can offer different kinds of measurement experiences.
- "show me" lengths
- How many smaller ones make a larger one?
- body measurements
- scavenger hunts
- seasonal measurements

–Jean Shaw and Mary Jo Puckett Cliatt
New Directions for Elementary School Mathematics
(NCTM Yearbook)

Children should explore the capacity of various containers by direct comparisons or by counting the number of scoops or cups required to fill each container. They should also experiment . . . and conjecture. . .
– Principles and Standards for School Mathematics (NCTM, 2000) page 104

MeasureWorks

Balances help students understand comparative weights and reinforce the concept of equality.

Scales permit students to assign numerical values to the weights of objects.
— *Principles and Standards for School Mathematics* (NCTM, 2000) page 104

"Measurement sense involves a knowledge of the measurement process."
—Jean Shaw and Mary Jo Puckett Cliatt
New Directions for Elementary School Mathematics
(NCTM Yearbook)

Mass/Weight

Mass refers to the amount of matter in an object. This measure remains the same regardless of the object's location. Weight refers to the gravitational force exerted by Earth or another celestial body on an object and eventually involves reading numeric scales. Students at all grade levels tend to use the terms *mass* and *weight* interchangeably. They need concrete examples in realistic contexts to help them distinguish the difference between an object's mass and its weight. Students need structured experiences in determining weights and sorting objects into different weight ranges (e.g., "These weigh a few ounces or grams." "These weigh more than 1 pound but less than 1 kg.") Younger students especially may be misled by the "look" of an object and mistakenly think that the larger the object, the more it weighs.

Temperature

refers to how hot or cold something is. It is relatively easy for students to learn to read a thermometer, but because of their inexperience, students do not distinguish well between temperatures that should be associated, for example, with "warm" vs. "cool." Students need to identify benchmark temperatures in both degrees Fahrenheit and Celsius for which they associate specific weather or cooking situations, such as the temperature on a hot summer day or the temperature of boiling water.

Enjoy using MeasureWorks as you watch your students learn and grow!

Carol Thornton, Ph.D.

References

Copely, Juanita V. (2000). *The Young Child and Mathematics*. Washington, DC: National Association for the Education of Young Students.

Kenney, Patricia Ann & Vicky L. Kouba. (1997). "What Do Students Know About Measurement?" In *Results from the Sixth Mathematics Assessment of the National Assessment of Educational Progress*, edited by Patricia Ann Kenney and Edward A. Silver, pp. 141-163. Reston, VA: National Council of Teachers of Mathematics.

Lehrer, Richard. (forthcoming). "Developing Understanding of Measurement." In *A Research Companion to NCTM's Standards*, edited by Jeremy Kilpatrick et al. Reston, VA: National Council of Teachers of Mathematics.

National Council of Teachers of Mathematics (2000). *Principles and Standards for School Mathematics*. Reston, VA: The Council.

Shaw, Jean and Mary Jo Cliatt. (1989). In *New Directions for Elementary School Mathematics*. Reston, VA: National Council of Teachers of Mathematics.

Managing Behavior to Support Hands-on Learning

Hands-on learning helps students understand abstract concepts and practice concrete skills. To make hands-on learning successful, students need to know <u>and agree to</u> rules of behavior.

Take the time to have students develop rules of behavior for the class.

The steps outlined below can help. At each step, be sure to write all of the ideas students suggest.

>Ask: What rules can help us learn through activities?
>
>Write each idea for rules in the WHAT column.
>
>Then ask: How can you help keep each rule?
>
>Write each idea for how to keep the rules in the HOW column.

Post the rules at each activity center and at the front of the room.

>REMEMBER: When students participate in making the rules, they will find it easier to live by them.

Now you can help students practice good activity skills. Remind them of the rules as needed.

OUR CLASSROOM RULES

#	WHAT	HOW
1.	Show respect for others.	Use quiet voices. Use polite language.
2.	Work together.	Give everybody a turn. Say what you need. Listen to what others say. Ask for help if you need it.
3.	Be fair.	Share.
4.	Respect the classroom.	Use materials carefully. Clean up after yourself.

MeasureWorks™

Scope and Sequence

Grade	1	2	3	4	5
MEASUREMENT CONCEPTS					
Choose the appropriate tool	✓	✓	✓	✓	✓
Choose the appropriate unit	✓	✓	✓	✓	✓
Estimate using benchmarks	✓	✓	✓	✓	✓
TIME CONCEPTS					
First/Next/Last; Before/After	✓				
Yesterday, Today, Tomorrow	✓				
Morning, Noon, Night	✓				
A.M. or P.M.	✓	✓	✓	✓	✓
Identify parts of a clock	✓	✓			
Know calendar concepts					
days of week	✓	✓	✓	✓	✓
months of year	✓	✓	✓	✓	✓
decade and century			✓	✓	✓
Identify equivalent units	✓	✓	✓	✓	✓
days/weeks	✓	✓			
minutes/hours		✓	✓	✓	✓
Compare by converting units			✓	✓	✓
Read/make a schedule	✓	✓	✓	✓	✓
TELL TIME					
Tell time to nearest					
hour or half-hour	✓	✓			
15 minutes		✓	✓	✓	✓
5 minutes			✓	✓	✓
1 minute			✓	✓	✓
Read and write time					
using digital/analog clocks	✓	✓	✓	✓	✓
Tell elapsed time		✓	✓	✓	✓
estimate duration			✓	✓	✓
estimate enough time			✓	✓	✓
predict early/late/on time			✓	✓	✓
calculate how much earlier/later				✓	✓
by comparing 2 clocks					✓

Grade	1	2	3	4	5
LENGTH CONCEPTS					
Compare (more/less; –er/–est)	✓	✓			
Identify everyday benchmarks	✓	✓	✓	✓	✓
Choose appropriate unit and tool	✓	✓	✓	✓	✓
Use nonstandard units	✓				
Estimate customary benchmarks		✓	✓	✓	✓
Estimate metric benchmarks		✓	✓	✓	✓
MEASURE LENGTH					
Estimate with nonstandard units	✓	✓			
Estimate with standard units	✓	✓	✓	✓	✓
Choose the appropriate unit			✓	✓	✓
Measure to nearest					
inch or half-inch	✓	✓	✓	✓	✓
foot		✓	✓	✓	✓
yard		✓	✓	✓	✓
mile			✓	✓	✓
Measure to nearest					
centimeter	✓	✓	✓	✓	✓
millimeter			✓	✓	✓
meter		✓	✓	✓	✓
kilometer			✓	✓	✓
Convert among customary units			✓	✓	✓
Convert among metric units			✓	✓	✓
Estimate between systems					✓
AREA					
Estimate area		✓	✓	✓	✓
Measure area with nonstandard units		✓	✓	✓	✓
Find area in metric units					
centimeters		✓	✓	✓	✓
meters			✓	✓	✓
Find area in customary units			✓	✓	✓
Calculate area using scale drawings					✓

MeasureWorks™

Scope and Sequence

Grade	1	2	3	4	5
PERIMETER					
Estimate perimeter		✓	✓	✓	✓
Measure perimeter in customary units					
inches		✓	✓	✓	✓
feet or yards			✓	✓	✓
Measure perimeter in metric units					
centimeters		✓	✓	✓	✓
meters		✓	✓	✓	✓
Calculate perimeter using scale drawings					
CAPACITY/VOLUME					
Measure capacity in nonstandard units	✓	✓	✓	✓	
Estimate capacity with nonstandard and standard units	✓	✓	✓	✓	✓
Compare capacity to one cup	✓	✓	✓	✓	✓
Measure capacity with customary units					
cup		✓	✓	✓	✓
pint		✓	✓	✓	✓
quart		✓	✓	✓	✓
gallon			✓	✓	✓
Measure capacity with metric units					
liters		✓	✓	✓	✓
milliliters		✓	✓	✓	✓
Convert among customary units			✓	✓	✓
Convert among metric units				✓	✓
Estimate between systems					
Measure volume in nonstandard units	✓	✓	✓	✓	
Measure volume using scale drawings					
WEIGHT					
Compare to one pound	✓	✓	✓	✓	
Estimate					
with nonstandard units	✓	✓	✓	✓	✓
with standard units	✓	✓	✓	✓	✓

Grade	1	2	3	4	5
WEIGHT (continued)					
Measure with customary units					
ounces	✓	✓	✓	✓	✓
pounds	✓	✓	✓	✓	✓
tons			✓	✓	✓
Measure with metric units					
grams	✓	✓	✓	✓	✓
kilograms		✓	✓	✓	✓
milligrams					✓
Convert among customary units			✓	✓	✓
Convert among metric units			✓	✓	✓
Estimate between systems					
TEMPERATURE					
Develop benchmark vocabulary (cool, warm, hot, cold, freezing)	✓	✓	✓	✓	✓
Identify benchmark temperatures (freezing point, boiling point)	✓	✓	✓	✓	✓
Read and use a thermometer					
Fahrenheit	✓	✓	✓	✓	✓
Celsius	✓	✓	✓	✓	✓
Estimate temperature	✓	✓	✓	✓	✓
Measure temperature	✓	✓	✓	✓	✓
Record temperature		✓			✓
ANGLES					
See angles as fractions of circles					
Estimate acute, right, obtuse			✓	✓	✓
Identify right and straight			✓	✓	✓
Compare to find greater or less than 90°				✓	✓
Identify and use benchmarks 30°, 45°, 60°, 90°					
Measure angles				✓	✓
Draw angles of given measures					

MeasureWorks™ • Grade 1

Measure Works™
Time

Unit Introduction

In this unit, students

☞ Tell the date using a calendar.

☞ Master calendar skills.

☞ Make a model of a clock.

☞ Estimate time to the hour and half-hour.

☞ Show time to the hour and half-hour.

☞ Tell time to the hour and half-hour.

KWL

Use a KWL chart to activate prior knowledge and set learning goals as a class. A reproducible KWL chart is provided on page BLM • 7.

Have students keep the KWL chart in their math folders and add to it as they work through this unit.

Assessment

A unit test in multiple-choice format is provided on page Assessment • 1.

Games for Practice and Review

Use the MeasureWorks Game Board to reinforce learning. Game rules begin on page BLM • 13.

Focus on Vocabulary

after (p. T-5)	December (p. T-3)	January (p. T-3)	months (p. T-3)	short hand (p. T-7)
analog (p. T-9)	digital (p. T-9)	July (p. T-3)	November (p. T-3)	Sunday (p. T-2)
April (p. T-3)	estimate (p. T-5)	June (p. T-3)	o'clock (p. T-5)	Thursday (p. T-2)
August (p. T-3)	February (p. T-3)	long hand (p. T-7)	October (p. T-3)	time (p. T-1)
before (p. T-5)	Friday (p. T-2)	March (p. T-3)	sand timer (p. T-11)	Tuesday (p. T-2)
between (p. T-6)	half-hour (p. T-8)	May (p. T-3)	Saturday (p. T-2)	Wednesday (p. T-2)
calendar (p. T-1)	hour (p. T-8)	minute hand (p. T-5)	schedule (p. T-12)	week (p. T-2)
clock (p. T-1)	hour hand (p. T-5)	Monday (p. T-2)	September (p. T-3)	

Write each vocabulary word on an index card. Display the cards in a pocket chart.
To reinforce literacy skills, have volunteers identify a selection of words as you say them.
To develop alphabetizing skills, make a selection of six related words. Have students help you put them in alphabetical order in a pocket chart.

Heads Up!

Calendar skills develop over time. The daily calendar activity is essential to mastering this skill. Daily practice builds both verbal and auditory skills. Analog clocks can be confusing to many students. Be sure to give students many opportunities to estimate time on a one-handed clock. Have students estimate time using only the hour hand to develop visual estimation skills. This practice will also help them make sense of the minute hand.

Book Nook

Chicken Soup with Rice

by Maurice Sendak

1962: Scholastic, Inc.

Catchy rhymes describe each month of the year and activities that go with them.

Time

Manipulatives

Pages	Learning Goals	Demonstration clock	Student clocks
T-1–1	Review, sort, and classify time words.		
T-2–2	Name and order the days of the week.		
T-3–3	Name and order months of the year. Sort and name months by season.		
T-4–4	Review months of the year and days of the week. Make a daily calendar.		
T-5–5	Tell time to the hour.	✓	✓
T-6–6	Estimate time before and after the hour.	✓	
T-7–7	Show and write time to the half-hour.	✓	✓
T-8–8	Tell time to the half-hour. Show time to the half-hour on an analog clock.	✓	✓
T-9–9	Match times on digital and analog clocks.	✓	✓
T-10–10	Understand different ways of expressing time to the half-hour. Match times on digital and analog clocks and in words.	✓	✓
T-11–11	Estimate short lengths of time. Understand the duration of a minute.		
T-12–12	Read a daily schedule. Understand that a schedule shows events in order.		
T-13–13	Make a schedule of after-school activities.		

MeasureWorks™ • Grade 1

Time • T-1B

Review Time Vocabulary

Planning Your Time
Intro & Demo	Activity	Sum It Up
15 min	15 min	10 min

Objective
Review, sort, and classify time words.

Materials
- Scissors
- Glue or paste
- Measurement Masters 1–2
- Tagboard strips
- Pocket chart (optional)

Grouping
Whole class, then individuals

Open It Up

Teaching Tip: Have students sit in a semi-circle for this activity.

Say: Today we are going to talk about **time.**

Ask: What words do you know that tell about time? [Samples: clock, calendar, o'clock, early, late]

Make a list of students' words, whether they are right or wrong, on the board.

Say: Some of these words have to do with **calendar** time. Some of these words have to do with **clock** time.

Read each word on the list. Have volunteers tell you whether it relates to the calendar or the clock.

Demonstrate & Discuss

Prepare ahead: Write the words *week, morning, evening, afternoon,* and the days of the week on tagboard strips.

Teaching Tip: Integrate literacy skills with this activity. As you read each word aloud, track the print and emphasize the initial letter sound.

Say: Let's read a word. Let's decide if it is a calendar word or a clock word.

Hold up a tagboard strip with the word *week*. Read the word, tracking the print and emphasizing the initial letter sound. Then have students read the word with you.

Ask: Is this a calendar word? [yes] How do you know? [Sample: A calendar shows weeks and days.]

Repeat with *morning.*

Ask: Is this a calendar word? [no]

Model how to organize for sorting. Set calendar words on the chalk ledge or a pocket chart. Set clock words aside.

Student Activity

Prepare ahead: Each student will need a pair of scissors, glue or paste, and copies of Measurement Masters 1–2.

Teaching Tip: Review safety rules related to scissors before distributing materials to students.

Read the directions on the student page aloud to students. Students cut apart rectangles that contain time words or copy the words on index cards. They sort and classify words into two groups—words that tell about clock time and words that tell about calendar time.

Then they make collages to share with their families.

Teaching Tip: Allow time for students to cut out the rectangles on their own.

Informal Assessment

Help students sort the words one at a time. Have them read each word.

Ask: Is this a clock word? How do you know? [Answers will vary.] /INFER/
What do you do with the word if it is a clock word? [Paste it on the clock sheet.] /FOLLOW DIRECTIONS/

Sum It Up

Say: Today we read words that tell about time. We sorted clock words and calendar words.

Ask: How do you know if a word is a clock word? a calendar word? [Samples: Clock words tell about what we see on the clock. Clock words tell about minutes and hours. Calendar words tell about days, weeks, and months.] /GENERALIZE/

Extension

Invite students to tell or write stories that use one or more of the vocabulary words. Have them find the words they used on their clock or calendar sheet and color them. Alternatively, they may draw pictures that illustrate one or more of the words.

Name _____

Clock and Calendar

Try This

- Cut out the boxes below.
- Sort the clock words.
- Paste them on the clock.
- Sort the calendar words.
- Paste them on the calendar.

Make two piles. Put clock words in one pile. Put calendar words in a different pile.

o'clock	afternoon	tomorrow	hour
now	late	night	minute
early	yesterday	later	month
soon	day	now	long hand
morning	week	year	short hand

MeasureWorks™ • Grade 1

Time • 1

Days-of-the-Week Memory

Planning Your Time
- Intro & Demo: 5 min
- Activity: 20 min
- Sum It Up: 10 min

Objective
Name and order the days of the week.

Materials
- Days of the week written on tagboard strips
- Measurement Master 3
- Scissors (or game cards prepared ahead)

Grouping
Whole class, then pairs

Open It Up

Teaching Tip: Reinforce literacy skills as you review the days of the week. Have students identify the initial consonant sounds to help recognize the names of days of the week.

Display seven tagboard strips, each showing a day of the **week.**

Invite students to help you put them in order.

Say: Today is **Tuesday.** Which says Tuesday? What day comes after Tuesday? [**Wednesday**] What day comes before **Monday?** [**Sunday**]

Continue in the same way until the days are in order.

Demonstrate & Discuss

Explain that today students will play a memory game called "Days of the Week."

Demonstrate how to play. One partner turns up two cards. One of the cards must be the "Today" card. The second card must match today's day. If neither card is the Today card, turn both facedown. The second partner turns up two new cards. If neither card is the Today card, turn both facedown. Continue until one person has matched "Today" with the name of the day.

Then take turns to match each day with its description. When five pairs have been matched, one partner tells one day that is missing and turns up one card. If the guess is correct, the player puts that card in correct order. The other partner guesses which day is missing and turns up the last card.

Student Activity

Prepare ahead: Each pair will need a pair of scissors and a copy of Measurement Master 3.

Read the directions on the student page aloud to students. Students practice ordering the days of the week by matching day cards with the card that tells the correct relationship to "Today."

The activity involves both visual memory and practice naming the days before and after a given day of the week.

Teaching Tip: Remind students to carefully put the cards back in the same place when they do not have a match. This will help students remember where given cards are when they have enough information to make a match.

Informal Assessment

Encourage students to read the print on the first card they turn over. Then, BEFORE they choose their second card, have them predict what the next card should say in order to get a match.

Ask: Which card should go with "Tomorrow?" How do you know? [Sample: Wednesday, because today is Tuesday and Wednesday comes after Tuesday.] /PREDICT/

Sum It Up

Say: Today we reviewed the names of the days of the week. We played a game to practice naming the days of the week in order.

Ask: What day comes before Tuesday? [Monday] What day comes after Sunday? [Monday] What day comes between Thursday and Saturday? [Friday] /RECALL/

Have students recite the days of the week orally, beginning with Sunday.

Daily Calendar Time

Have students draw a picture of the weather for today and label it "Today." Then have them predict what the weather will be tomorrow. Have them draw their prediction and label it "Tomorrow."

Name _____

Remember the Days!

Measure Works™

Try This

- Work with a partner.
- Cut out one set of game cards.
- Lay the cards facedown.
- Make three rows. Put four cards in each row.
- Take turns.

Try to remember where each card you have turned over is.

Player 1:

- Turn up two cards. Do they match?
- If they match, place them faceup side by side.
- If they do not match, turn them back facedown.

Player 2:

- Repeat.

MeasureWorks™ • Grade 1   Time • 2

Name and Order Months of the Year

Planning Your Time
Intro & Demo: 15 min
Activity: 20 min
Sum It Up: 10 min

Objective
Name and order months of the year.
Sort and name months by season.

Materials
- Months of the year written on tagboard strips
- 11" × 17" construction paper
- Scissors

Grouping
Whole class, then individuals

Open It Up

Explain that today students will review **months** of the year. Display the months of the year in random order.

Ask: Which is the first month of the year? [**January**] Point out that on January 1st each year, we start a new year.

Ask: Which month comes after January? [**February**]

Have students use initial consonants to identify the strip that names February. Place it below January.

Continue in the same way, inviting students to help you put the tagboard strips in order.

Demonstrate & Discuss

Have students chorally recite the months in order, using the tagboard strip display.

Teaching Tip: Reinforce literacy skills by exaggerating the initial letter sound for each month.

Review the seasons of the year. Ask students to sort the months of the year into four seasons.

Student Activity

Prepare ahead: Each student will need a sheet of 11" × 17" construction paper and a pair of scissors. Read the directions on the student page aloud to students. Students make a Seasons book. They use cutouts to classify the months by season. Then they order the months within each season. They fold a sheet of construction paper in half to make a four-page book and paste the months in order on pages for the appropriate seasons. Encourage students to decorate each seasonal page with symbols of the season.

Teaching Tip: Help students put the months in the correct order before they paste them into their Seasons Book.

Say: Winter begins in December. Which strip names December? [Help students identify the month that begins with D.] Which month comes next? [January]

Informal Assessment

Check that students are putting the months in the correct order.

Ask: Which is the first month of the year? [January] /RECALL FACTS, SEQUENCE/
Which month comes before June? [May] /SEQUENCE/
Which month comes after March? [April] /SEQUENCE/
Which month comes before October? [September] /SEQUENCE/

Sum It Up

Say: Today, we practiced saying the months of the year. We sorted the months into seasons.

Have students recite the months of the year in order.

Ask: How many months are in one year? [12]

Which is the first month of the calendar year? [January] /SEQUENCE/

Which is the last month of the year? [December] /SEQUENCE/

Book Nook

Chicken Soup with Rice by Maurice Sendak (Scholastic, 1962)

Talk about how each verse helps identify the month of the year. Have students make a list of clues for each month.

Now, without reading the name of the month, read the verse that describes a month. Have students guess which month goes with the description. Repeat as time allows.

Name _____

My Seasons Book

Hint: The four seasons are fall, winter, spring, and summer.

Try This
- Cut the cards apart.
- Put the months in order.
- Sort by season.
- Fold a piece of paper in half.
- Paste three months for each season.
- Decorate your Seasons Book.

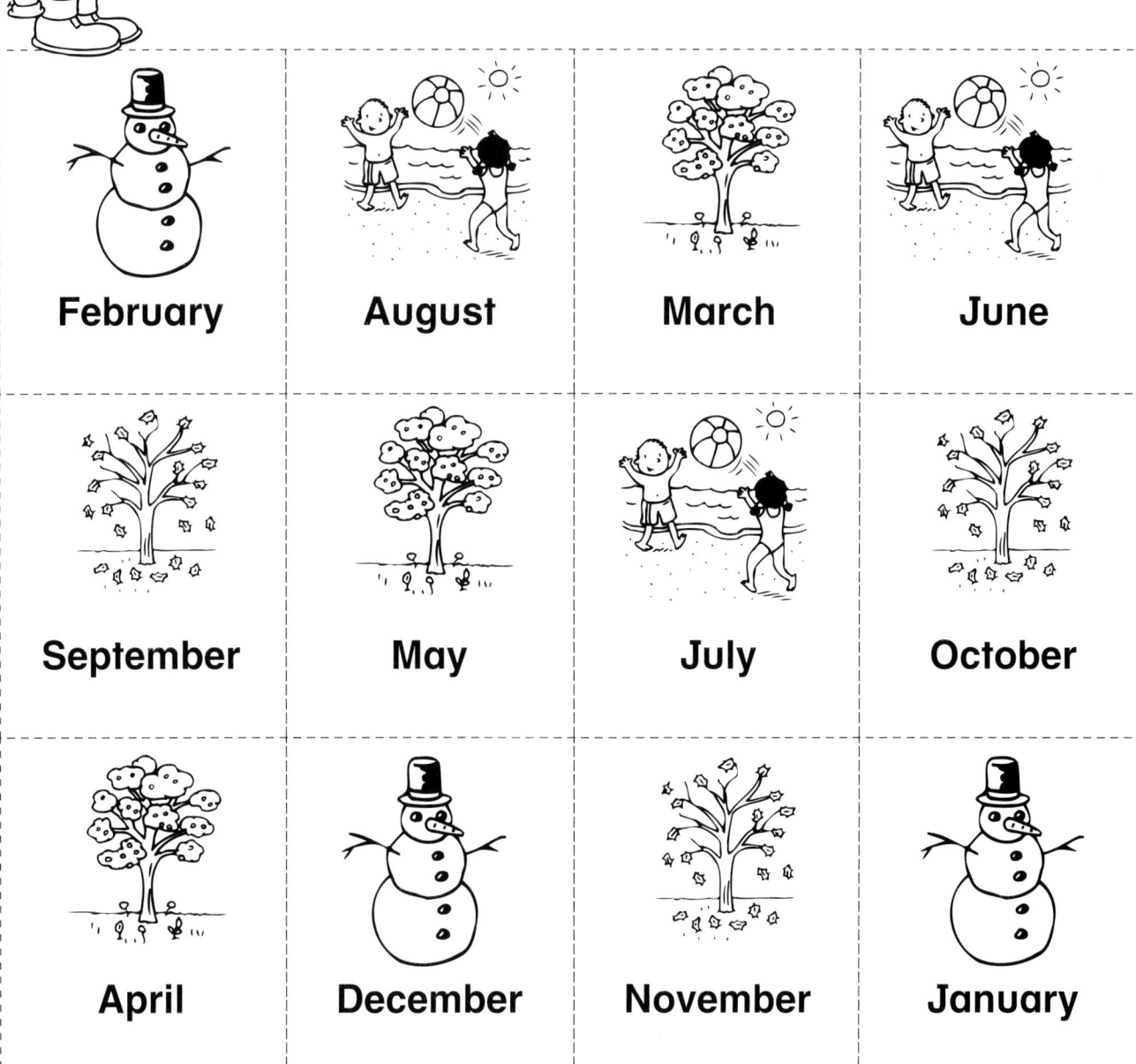

February	**August**	**March**	**June**
September	**May**	**July**	**October**
April	**December**	**November**	**January**

MeasureWorks™ • Grade 1   Time • 3

Daily Calendar Activity

Planning Your Time
Intro & Demo	Activity	Sum It Up
10 min	10 min	10 min

Objective

Review months of the year and days of the week. Make a daily calendar.

Materials

- Measurement Master 4
- Glue or paste
- Scissors

Measurement Master 4

Grouping

Whole class, then individuals

Open It Up

Teaching Tip: This activity is designed to become part of your daily routine. Incorporate it into circle time or make it a part of taking attendance each day.

Display the names of the months of the year on tagboard strips.

Guide a group reading of the months. Point to each month as you say it with the students.

Repeat with the days of the week.

Then distribute calendars. Explain that each day, students will add to their calendar. At the end of the month, they can take the calendar home to share with their families.

Demonstrate & Discuss

Prepare ahead: Make a calendar bulletin board.

Ask: What month is it? [Sample: September]
What letter does September begin with? [*S*]
Who can find the label that says September?

Have a volunteer identify the correct strip and bring it to you to place on the calendar bulletin board.

Ask: What day of the week is it? [Sample: Monday]
Who can find the label that says Monday?

Have another volunteer identify the correct strip and bring it to you to place on the calendar bulletin board.

Ask: What day comes before Monday? [Sunday]
What day comes after Monday? [Tuesday]

Student Activity

Prepare ahead: Each student will need a copy of Measurement Master 4, glue or paste, and a pair of scissors.

Read the directions on the student page aloud to students. Students work individually. They trace the letters to spell each month and read the list aloud. Then they cut out the label for the current month and paste it on their calendars.

Introduce the daily calendar routine. Each day, students identify the date, the day of the week, the month, and the year.

Then they write the day's information on their calendar. Encourage them to draw the sun a cloud, a rain drop, or a snow flake to record the weather.

Students complete the following recitation as a class each day to practice calendar skills.

"Today is [day], the [date] of [month] 2003."
"Yesterday was [day], the [date] of [month] 2003."
"Tomorrow will be [day], the [date] of [month] 2003."

Informal Assessment

Ask: What season is April in? [spring] /RECALL/

What month was it last month? [Answers will vary.]
What month will it be next month? [Answers will vary.]
/RECALL, SEQUENCE/
Which months start with *M*? [March, May] /OBSERVE/

Sum It Up

Say: Today we named the months of the year in order. We named the days of the week in order. We wrote in our calendar.

Ask: What month comes after December? [January]
What month comes before July? [June]
What day comes before Thursday? [Wednesday]
What day comes after Tuesday? [Wednesday]
/RECALL, COMPREHENSION/

Music Connection

Chicken Soup with Rice by Maurice Sendak

Help students learn one or more verses of this playful song. You may want to play this tape on a regular basis to signal clean-up time. At the end of the unit, stage a performance of Chicken Soup with Rice, having pairs of students perform the verse for each month.

Time • T-4

MeasureWorks™ • Grade 1

Name _____

My Calendar

Try This

- Read the name of each month.
- Trace the letters.
- Read the list aloud.
- What month is it now?
- Cut it out.
- Paste it on your calendar.
- Draw a star by today's date.

Draw a picture to show the weather for each day.

January	February
March	April
May	June
July	August
September	October
November	December

MeasureWorks™ • Grade 1   Time • 4

Estimate Time to the Hour

Planning Your Time
Intro & Demo	Activity	Sum It Up
20 min	15 min	5 min

Objective
Tell time to the hour.

Materials
- Student clocks
- Demonstration clock
- Overhead transparency (see Demonstrate and Discuss)
- Paper plates or Measurement Master 5 copied onto tagboard
- Scissors
- Brads

Measurement Master 5

Grouping
Whole class, then individuals

Open It Up
Display the demonstration clock.
Ask: What do you know about the hands of a clock? [Sample: There are two (or three) of them. They move. One hand is shorter than the other.]
Ask How do the hands on a clock help us tell time? [Sample: The short hand, or **hour hand,** shows the hour. The long hand, or **minute hand,** shows the minutes.]
Distribute a paper plate or a copy of Measurement Master 5 to each student. Have students write the numbers from 1 to 12 in order around the clock. Then have students cut out the hour hand and fasten it to the center of the clock with a brad.

Demonstrate & Discuss
Draw a one-handed clock (hour hand only) on an overhead transparency. Draw the hour hand pointing to the 3.
Ask: Which hand is missing from the clocks we made? [the minute hand]
Say: Today we will learn to **estimate** the time using just the hour hand.
Tell students that when the hour hand points to a number on the clock, it tells time to the hour. When the hour hand points to a space between two numbers on the clock, it tells the time between the hours.
Ask: Is the hour hand **on** the 3, **before** the 3, or **after** the 3? [on the 3] What time could it be? [about 3 **o'clock**]
Erase the hand and draw a new one between the 3 and the 4.
Ask: What time could it be now? [between 3 and 4 o'clock]
Say: So the time is between 3 and 4 o'clock.

Time • T-5

Student Activity
Prepare ahead: Each student will need a student clock.
Point out that the hour hand is the shorter of the two clock hands. Explain that the hands on students' hand-made clocks are hour hands.
Read the directions on the student page aloud to students. Students look at the first picture on the recording sheet.
Ask: What number is the hour hand pointing to? [12] What time can it be? [about 12 o'clock]
Say: Show me the same time on your student clock. Check students' clocks.
Then demonstrate how to record the time on the recording sheet. Repeat this process with each example.

Informal Assessment
Help students practice the vocabulary by having them read each time in unison.
Ask: Is the hour hand on a number or between two numbers? [Answers will vary.] /DESCRIBE/

Sum It Up
Say: Today we learned how to tell time to the hour. The hour hand tells us time to the hour.
Ask: How do you know when it is about 10 o'clock? [Sample: The little hand on the clock is pointing to the 10.] /SUMMARIZE/
How do you know when it is between 2 and 3 o'clock? [Sample: The short, hour hand has passed the 2 and not yet reached the 3.] /GENERALIZE/

Extension
Have students work in pairs to play guessing games with their clocks. One student shows a time on his or her clock, but does not show it to the other player. The other player asks yes-or-no questions to guess the time: Is the hour hand after the 6? on the 7? between 7 and 8? When a player feels ready to guess the actual time, he or she asks: Is your time between 8 and 9 o'clock?

MeasureWorks™ • Grade 1

Name _____

It's About Time

MeasureWorks

Try This

- Show the same time on your 🕗 .
- Say the time.
- Write the time.

I can estimate time to the hour. I look at the hour hand.

①  About _____ o'clock

② About _____ o'clock

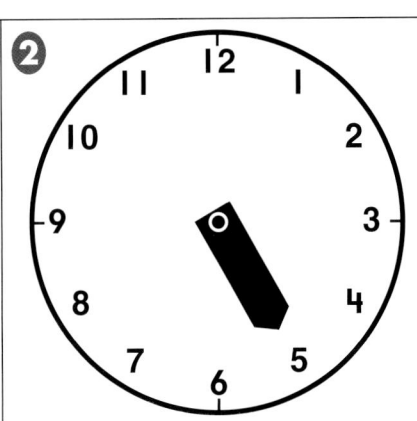

③ About _____ o'clock

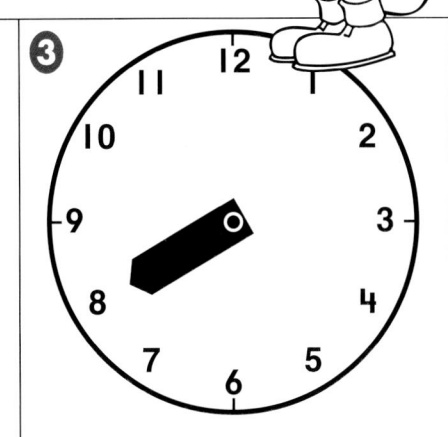

④ About _____ o'clock

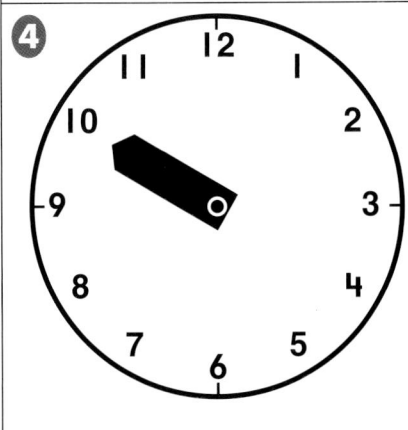

⑤ About _____ o'clock

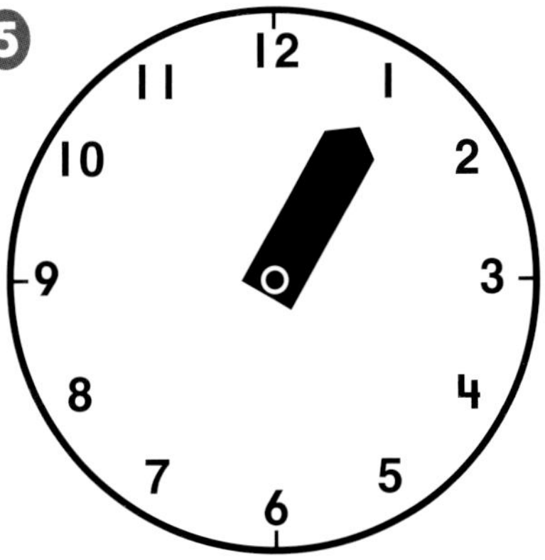

⑥ About _____ o'clock

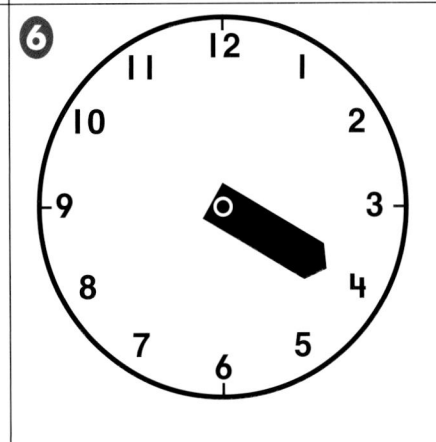

MeasureWorks™ • Grade 1   Time • 5

Tell Time Before and After the Hour

Planning Your Time
Intro & Demo	Activity	Sum It Up
15 min	15 min	10 min

Objective
Estimate time before and after the hour.

Materials
- Demonstration clock
- Student-made one-handed clocks (see p. Time • T-3)
- Overhead transparency

Grouping
Whole class, then individuals

Open It Up

Draw a one-handed clock on an overhead transparency. Draw the hour hand on the 4.

Ask: Where is the hour hand? Is it on a number or **between** two numbers? [on the 4]

Now erase the hour hand and draw a new one to point between 4 and 5.

Ask: Where is the hour hand now? Is it on a number or between two numbers? [between 4 and 5] Is it before the 4 or after the 4? [after the 4] Is it before the 5 or after the 5? [before the 5]

Say: The hour hand is between the 4 and the 5. So the time on this clock is between 4 o'clock and 5 o'clock. This means that it is 4 o'clock plus some minutes.

Ask: When will it be 5 o'clock plus some minutes? [when the hand is between the 5 and the 6]

Demonstrate & Discuss

Have students set the hour hand on their individual one-handed clocks to a time of their choice.

Ask for a volunteer to share his or her clock with the class. Have the volunteer **ask:** "Where is the hour hand?"

Have a classmate describe the hour hand.

Encourage students to use the most appropriate sentence frame:

I see that the hour hand is _____.
[Sample: on the 3; between the 7 and the 8]

The time is _____
[Sample: 3 o'clock; after 7 o'clock; before 8 o'clock]

Repeat with other volunteers as time allows.

Student Activity

Prepare ahead: Each student will need a student-made one-handed clock (see previous lesson).

Read the directions on the student page aloud to students. Students use their one-handed clocks to show time before and after the hour. They estimate the time and complete sentences using the words *on, between, before,* and *after.*

Informal Assessment

Help students use the lesson vocabulary as they work.

Ask: Where is the hour hand? [Sample: The hour hand is between 6 and 7.] /OBSERVE, DESCRIBE/

Students read the sentences for that clock. Have them track the print as they read.

Sum It Up

Say: Today we learned to tell time before and after the hour. When the hour hand points to a space between two numbers on the clock, it tells the time between the hours.

Ask: How do you know if the time is before or after the hour? [The hour hand is between two numbers.] /GENERALIZE/

Extension
Use Measurement Master 5 to practice showing clock time.

Over the course of the day, have students stop and look at the classroom clock. Have them draw the position of the hour hand. Then have them estimate the time. Ask: "Is it before [hour] o'clock, about [hour] o'clock, or after [hour] o'clock?"

Name _____

Color, Show, Write

Measure Works

Try This

- Color the hour hand red.
- Show the same time on your clock.
- Complete each sentence.

Remember: The hour hand is short and the minute hand is long.

Color the hour hand red.	Complete each sentence. Then write <u>before</u> or <u>after</u>.
❶ (clock showing hour hand between 3 and 4, pointing down toward 6)	The time is between _____ o'clock and _____ o'clock. It is _____ 3 o'clock. It is _____ 4 o'clock.
❷ (clock showing hour hand between 9 and 10, pointing down toward 6)	The time is between _____ o'clock and _____ o'clock. It is _____ 10 o'clock. It is _____ 9 o'clock.
❸ (clock showing hour hand between 11 and 12, pointing toward 3)	The time is between _____ o'clock and _____ o'clock. It is _____ 11 o'clock. It is _____ 12 o'clock.
❹ (clock showing hour hand pointing between 7 and 8, toward 9)	The time is between _____ o'clock and _____ o'clock. It is _____ 8 o'clock. It is _____ 7 o'clock.

MeasureWorks™ • Grade 1   Time • 6

Explore the Minute Hand

Planning Your Time
Intro & Demo	Activity	Sum It Up
10 min	15 min	10 min

Objective
Show and write time to the half-hour.

Materials
- Student clocks
- Demonstration clock

Grouping
Whole class, then individuals

Open It Up

Distribute student clocks. Review time to the hour.

Say: The big red numbers show the hours.

Ask: What is the color red on your clocks? [the **short hand,** the hour hand, the numbers 1–12] What is the color blue on your clocks? [the **long hand,** minute hand, the tiny numbers]

Say: Let's read the blue numbers together. Start with the small 5 that is by the 1.

Skip count together by 5s to 60.

Say: There are 60 minutes in an hour. The small blue numbers tell us minutes past the hour.

Demonstrate & Discuss

Have students turn the minute hand to the 12. Then guide them to turn the minute hand in number order from 1 through 12.

Have students turn the minute hand to the 12.

Ask: Where is the hour hand now? [The hour hand is on one of the numbers.]

Say: Now slowly push the minute hand once around the clock.

Ask: What happened to the hour hand? [It moved from one number to the next.]

Now show 1:30 on the demonstration clock. Challenge students to turn the minute hand on their individual clocks until they show the same time. Point out that the minute hand moved halfway around the clock. Have students skip count by 5s to 30.

Say: There are 30 minutes in half an hour. Your clocks show 30 minutes past the hour, or 30 minutes past 1 o'clock.

Model how to record the time 1:30.

Student Activity

Prepare ahead: Each student will need a student clock.

Read the directions on the student page aloud to students. Read the following times one at a time: 3:00, 1:30, 8:30, 5:00, and 2:30. Students listen to each time and model it on their student clocks. Then they record the time by drawing the hour hand and the minute hand. They look at the chart to find which animal has playtime at each time, and they write the animal's name next to the clock.

Informal Assessment

Help students notice the relative positions of the minute hand and the hour hand.

Ask: Where is the hour hand when the minute hand is on the 12? [on a red number that tells the hour] /DESCRIBE/ Where is the minute hand when the time is 12:30? [on the little blue 30] /DESCRIBE/ How many minutes have passed when you move the minute hand from the 12 to the 6? [30 minutes] /INFER/ How does the hour hand move when you move the minute hand from 30 min to 45 min? [the hour hand moves closer to the next hour] /PREDICT, DESCRIBE/

Sum It Up

Say: Today we learned that there are 60 minutes in an hour. We learned that when the minute hand has gone around the clock once, the hour hand has moved one hour.

Ask: How many minutes are in one hour? [60 minutes] /RECALL/
How does the minute hand move on a clock? [once around the clock every hour] /DESCRIBE/

Extension

Explain that you are going to play an estimation game. Say: Close your eyes. When I say, "Start," begin to estimate. When you think one minute has passed, raise your hand.

Note at what interval each student estimates a minute. Say, "Now" when a minute has passed. Report whose estimates were close.

Repeat frequently throughout the day.

Time • T-7

MeasureWorks™ • Grade 1

Name _____

Who Has Playtime?

Try This

- Listen to your teacher.
- Show the same time on your clock.
- Draw the time.
- Use the chart. Who is playing?
- Write who has playtime.

Telling time uses the minute hand and the hour hand.

PLAYTIMES

Dan the dog:	1:30
Matt the cat:	2:30
Ron the rat:	5:00
Fran the frog:	8:30
Pat the bat:	3:00

Draw the time.	Who has playtime?
❶	_____
❷	_____
❸	_____
❹	_____
❺	_____

MeasureWorks™ • Grade 1   Time • 7

Tell Time to the Half-Hour

Planning Your Time

Intro & Demo	Activity	Sum It Up
15 min	15 min	10 min

Objective

Tell time to the half-hour.
Show time to the half-hour on an analog clock.

Materials

- Student clocks
- Demonstration clock

Grouping

Whole class

Open It Up

Gather students in a half-circle around the demonstration clock. Set the demonstration clock to show 1:00. **Say:** This clock shows 1 o'clock.

Ask: Where is the hour hand? [on the 1]
What blue number does the minute hand point to? [60]

Point out that when the minute hand is on the red 12, it is also on the blue 60. The time is exactly on the hour.

Direct students to set their student clocks to show 1:00.

Demonstrate & Discuss

Now slowly turn the minute hand to show 1:30. Discuss how the clock has changed.

Ask: Where is the hour hand now? [between the 1 and the 2] Where is the minute hand now? [on the 6; pointing to the blue 30]

Say: The clock now shows 1:30. That means 30 minutes after 1. We also say half past 1 because 30 minutes is half of one **hour**.

Have students move the minute hand on their clocks to show 1:30.

Give individual students the opportunity to say the time.

Ask: What time does your clock show? [1:30]

Repeat with other examples of time to the hour and the **half-hour** later. Be sure students model each time using their clocks.

Student Activity

Prepare ahead: List nine different hour and half-hour times.

Read the directions on the student page aloud to students. Read various times to the hour and to the half-hour in the form " ___ o'clock" and "30 minutes past [a given hour]."

Students listen and draw to show the time on an analog clock. When they finish, they say the time aloud.

Informal Assessment

As students work, remind them to think of the positions of both the minute and the hour hand.

Ask: Where do you draw the hour hand when the time is 2:30? [between the 2 and the 3] /DESCRIBE/

Where do you draw the minute hand when the time is 2:30? [on the 6] /DESCRIBE/

Sum It Up

Say: Today we learned to tell time to the half-hour.

Ask: How many minutes are in an hour? [60] /RECALL/
How many minutes are in half an hour? [30] /RECALL/

Extension

Invite students to set their clocks to a time of their choice. Then ask, "Whose clock shows a time that is later than 3:00?" Have those students display their clocks. Discuss how to know when it is later than 3:00.

Repeat the activity, asking "Whose clock shows a time that is earlier than 12:00?"

Repeat with other examples as time allows.

Name _____

Hour and Half-Hour

Try This
- Listen to your teacher.
- Draw the hands to show the time.
- Say the time.

Remember: There are 60 minutes in one hour. There are 30 minutes in each half-hour.

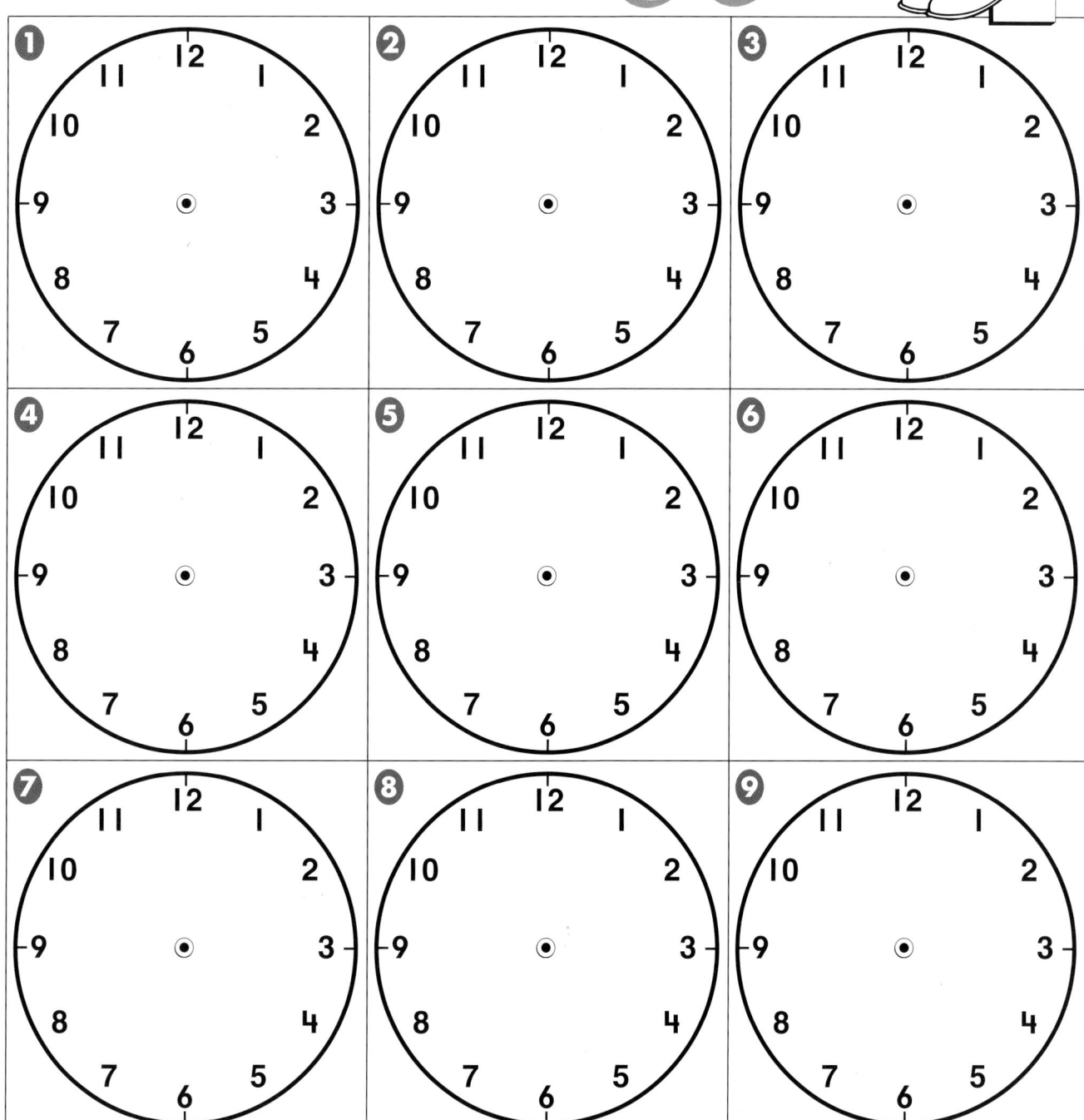

MeasureWorks™ • Grade 1 Time • 8

Read Digital and Analog Clocks

Planning Your Time

Intro & Demo	Activity	Sum It Up
15 min	15 min	10 min

Objective
Match times on digital and analog clocks.

Materials
- Student clocks
- Demonstration clock

Grouping
Whole class, then pairs

Open It Up

Show 4:00 on the demonstration clock.

Ask: Where is the hour hand? [on the 4]
Where is the minute hand? [Sample: on the 12]

Say: When the minute hand is pointing straight up to the 12, it shows zero minutes after the hour. We say the time is 4 o'clock.

Write *4 o'clock* and *4:00* on the board.

Say: Our clock shows 4 o'clock. We can write it as 4:00, which means zero minutes after the hour.

Repeat with other examples, such as 6:00 and 9:00.

Say: A clock that has hands and numbers going around it is called an **analog** clock.

Demonstrate & Discuss

Draw a digital clock face on the board. Tell students that a clock that shows time this way is called a **digital** clock.

Show 8:00 on the demonstration clock. Have a volunteer tell what time is shown. [8:00] Write 8:00 on the digital clock face.

Ask: What time did I write on the board? [8 o'clock] How many minutes past the hour does my clock show? [zero] Where is the minute hand? [on the 12]

Now turn the minute hand to show 8:30. Write 8:30 on the digital clock face.

Ask: What time does my clock show now? [30 minutes past 8 o'clock] What time did I write on the board? [8:30] How many minutes past the hour is 8:30? [30 minutes]

Student Activity

Prepare ahead: Each student will need a student clock.

Read the directions on the student page aloud to students. Students read time to the hour and half-hour in digital form. They write how many minutes past the hour each time shows. They they model the time on student clocks and draw the hands of an analog clock to show the time.

Informal Assessment

As students work, help them use their listening skills and time vocabulary.

Ask: Where is the minute hand when the time is 7 o'clock? [Sample: on the 12] /DESCRIBE/

Where is the hour hand? [on the 7] /DESCRIBE/

Sum It Up

Say: Today we learned that there are two ways to tell time. A digital clock tells time in digits. An analog clock tells time with hands on a clock face.

Ask: Where is the minute hand when the time is 7:30? [on the 6] /DESCRIBE/

Where is the hour hand when the time is 7:30? [between the 7 and the 8] /DESCRIBE/

Literature Connection

The Carrot Seed by Ruth Krauss (Harper and Row, 1945)

Read the story aloud. Then ask: What happened in this story? Make a list of responses. Invite the class to help put the events in story order. Reread the story. After each page, ask: What do you think will happen next? Open the book at random. Show the picture. Ask: What is happening here? What happened *before* this part?

Time • T-9

MeasureWorks™ • Grade 1

Name _____

Match the Clocks

MeasureWorks

Try This

At zero minutes past the hour, the minute hand is at 12.

- Read the time on the digital clock.
- Write the time.
- Show the same time on your 🕐.
- Draw the time.

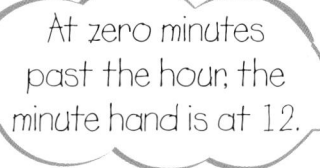

Read the time.	Write the time.	Draw the time.
❶ 7:30	_____ minutes past _____ o'clock	
❷ 5:00	_____ minutes past _____ o'clock	
❸ 11:30	_____ minutes past _____ o'clock	
❹ 2:30	_____ minutes past _____ o'clock	

MeasureWorks™ • Grade 1

Time • 9

Recognize Time to the Half-Hour

Planning Your Time
Intro & Demo: 10 min
Activity: 15 min
Sum It Up: 10 min

Objective
Understand different ways of expressing time to the half-hour.
Match times on digital and analog clocks and in words.

Materials
- Student clocks
- Demonstration clock
- Measurement Master 6
- Scissors

Grouping
Whole class, then pairs

Open It Up

Say: I'm thinking of a time. The hour hand is pointing to the 7 and the minute hand is pointing to the 12. What time am I thinking of? [7 o'clock]

Have students guess the time without using student clocks. Then have them show the time on student clocks.

Say: I'm thinking of another time. The hour hand is between the 5 and the 6 and the minute hand is on the 6. What time am I thinking of? [5:30]

Have students guess and show the time. Repeat as time allows.

Demonstrate & Discuss

Ask: How many minutes are in an hour? [60] How many minutes are in half an hour? [30]

Show 2:30 on the demonstration clock. Have students model the same time with student clocks.

Ask: How many ways can you think of to say or write this time? [2:30; two-thirty; 30 minutes past two; half past two]

Write students' responses on the board. Suggest any answers they miss.

Say: These are names that all describe the same time.

Ask: What kind of clock would show 2:30? [a digital clock] Why do we say it is half past two? [Samples: because the minute hand has gone halfway around; because 30 minutes is half of an hour]

Student Activity

Prepare ahead: Each pair will need one copy of Measurement Master 6, two student clocks, and a pair of scissors.

Read the directions on the student page aloud to students. Students work in pairs. Have students cut apart the cards on Measurement Master 6. They mix up the cards and spread them out facedown. Students take turns flipping over a card. If the card matches a time on an uncovered square on that player's student page, he or she puts the card on the square and shows the time on a student clock. Players continue to take turns until one player fills the chart. If a player turns over a card that he or she already has, he or she puts it back facedown in its place. The first player to fill the chart is the winner.

Informal Assessment

As students play, encourage them to read the cards and the times on the chart aloud to reinforce learning.

Ask: Which digital time is the same as half past three? [3:30] How do you know? [I know that half past means half an hour, or thirty minutes.] /RELATE/
What is another way to say five-thirty? [half past five; thirty minutes past five] /RECALL/

Sum It Up

Say: Today we learned to say the same time in several different ways.

Say: Turn over your charts. Look at your cards.

Ask: What is a different way to say or write each time? [Answers will vary.]

Have volunteers give two forms for each time on the cards. Show each time on the demonstration clock and have another volunteer describe the position of the hands.

Extension

Have students play "I'm Thinking of a Time" in pairs. Have one student show a time on a student clock, without showing the clock to the other player, and fill in these sentence frames:

This time is ____ minutes past the hour.

The hour hand is pointing _____.

The other partner guesses the time. Then they trade roles.

Name _____

Time Match-Up

Try This

- Play with a partner.
- Spread the cards out facedown.
- Take turns.
- Turn over a card.
- Look for the time on the chart.
- Put the card on the square that shows the same time.
- Show the same time on your clock.
- If you already have the time on the card, put it back in place.
- Now it is your partner's turn.

To win, be the first to fill your chart!

3:30	5:00	3:00
5:30	12:30	9:00
1:30	11:30	9:30

MeasureWorks™ • Grade 1   Time • 10

Estimate Elapsed Time

Planning Your Time
Intro & Demo	Activity	Sum It Up
15 min	20 min	5 min

Objective
Estimate short lengths of time.
Understand the duration of a minute.

Materials
- Wind-up toy or sand timer
- Clock(s) for timing 1 minute
- Optional: stop watch or timer

Grouping
Whole class, then pairs

Open It Up
Display the wind-up toy or **sand timer.** Have the class watch as you run the toy or timer once.

Ask: Do you think we can sing the alphabet song before the sand runs through/toy winds down? [Answers will vary.]

Have students vote silently with a thumbs up or a thumbs down. Count and record how many vote for each option.

Set the sand timer or toy and lead students as they sing the song together with their eyes closed. Check students' predictions.

Repeat with a different song, such as "Knick Knack Paddywack." Have students estimate how many verses they can sing before the sand runs through or the toy winds down.

Demonstrate & Discuss
Explain how to use a clock to measure a minute. If the clock has a second hand, direct students to say "go" when the hand is at 12 and "stop" when the hand returns to 12. Otherwise, direct students to say "go" when the minute hand reaches a minute mark and "stop" when the hand reaches the next minute mark.

If you have a stop watch or timer, show that one minute is the same on the clock as on the stop watch or timer. Have students watch the clock as you use the stop watch or timer to time one minute. If students will be operating the stop watch or timer, explain how to do this.

Student Activity
Prepare ahead: Each pair will need access to a clock, stop watch, or timer.

Read the directions on the student page aloud to students. Students work in pairs. In Column 1 on the student page, students list four activities they think will take one minute. Possible activities include the following: count to 100; walk down the hall and back; tie my shoes; or print my name five times.

Next, one student in the pair does the activity while the other student uses a clock (or timer or a stop watch) to time one minute. Students circle if the activity took less than a minute, more than a minute, or exactly a minute.

Teaching Tip: If students have difficulty thinking of activities to list, make a class list on the board and let students choose.

Informal Assessment
Encourage students to discuss their estimates and their results.

Ask: Which of your activities took less than (more than, exactly) one minute? [Answers will vary.] /DESCRIBE/

Did you usually guess the activities would take more time or less time than they really did? [Answers will vary.] /ANALYZE/

Sum It Up
Say: Today we estimated how much we could do in a certain length of time.

Ask: Did anyone list an activity that took exactly one minute? What was it? [Answers will vary.] /OBSERVE/
Can you explain how to use a clock (or timer or stop watch) to measure one minute? [Sample: Say "go" when the second hand is at 12, watch until it gets to the 12 again, and then say "stop."] /DESCRIBE/

Extension
Conduct an "Estimation Olympics." Have students estimate how long different athletic activities will take. For example, how long will it take to run around the school or to hop on one foot across the blacktop? Instead of rewarding the fastest students, award certificates to the students whose estimates were close to their actual times.

Name _____

Just a Minute

Measure Works™

Try This

- What do you think will take about 1 minute?
- List four different actions.
- Time each action.
- Circle the words that tell how long it took.

A minute is 60 seconds long.

Actions	Time it. Circle your answer.
❶	Less than 1 minute More than 1 minute Exactly 1 minute
❷	Less than 1 minute More than 1 minute Exactly 1 minute
❸	Less than 1 minute More than 1 minute Exactly 1 minute
❹	Less than 1 minute More than 1 minute Exactly 1 minute

MeasureWorks™ • Grade 1   Time • 11

Use a Schedule

Planning Your Time
Intro & Demo	Activity	Sum It Up
15 min	10 min	5 min

Objective
Read a daily schedule.
Understand that a schedule shows events in order.

Materials
- Mixed-up classroom schedule on poster board (see Open It Up)
- Usual classroom schedule on poster board

Grouping
Whole class, then individuals

Open It Up

Ask: What is a **schedule**? [Sample: a list of things to do and the times]

What kinds of schedules do you know about? [Samples: our daily classroom schedule, soccer practice schedule]

Tell students that you made a new classroom schedule and would like to know what they think of it. Display a mixed-up schedule, for example:

9:00 – lunch	10:00 – recess
11:30 – reading	1:00 – clean up to go home
1:30 – art	2:30 – math
3:00 – welcome	

Demonstrate & Discuss

Discuss the schedule.

Ask: What is the first activity on this schedule? [lunch] What time is recess? [10:00] What is on the schedule after cleanup time? [art] Is lunch at a good time? [No, it is first thing in the morning.]
Is recess before or after reading class? [before] What time is the last activity? [3:00]
How does this schedule differ from our usual schedule? [Sample: Reading is usually first.] Is this a good schedule? Why or why not? [No; the order doesn't make sense.]

Show students your usual classroom schedule. Ask questions about times that various activities are scheduled. Have students compare the clock and schedule to see if you are **on schedule** today.

Student Activity

Read the directions on the student page aloud to students. As a class, read the schedule for Camp Sunny, discussing words that may be unfamiliar to the students. Explain that students should use information from the schedule to answer the questions about camp activities. Students fill in the blanks to answer the questions.

Informal Assessment

As students work, encourage them to talk about the times and activities on the schedule.

Ask: Where will you look to find the answer to the questions? [on the schedule] /OBSERVE/

If you were going to Camp Sunny, what activity would you like best and when is it scheduled? [Answers will vary.] /DESCRIBE/

Which activities are before (after) lunch? [Sample: a hike] /SEQUENCE/

Sum It Up

Say: Today we practiced reading daily schedules.

Ask: How are schedules useful? [Samples: Schedules help us know what to expect and when. A schedule helps everyone to cooperate and work together.] /GENERALIZE/

Bulletin Board Idea

Post the daily schedule for your classroom on a bulletin board. Take photos of students as they participate in the different activities listed on the schedule. Show each photo to the class, and discuss the activity and its scheduled time. Then tack the photos around the schedule to make a border for the bulletin board.

Name _____

Camp Sunny

Measure Works

Try This

- Rod gets to spend the day at Camp Sunny.
- Read the camp schedule.
- Answer the questions. Fill in the blanks.

CAMP SUNNY SCHEDULE

9:00	Hike
12:00	Lunch
1:30	Nature walk
2:30	Swim
5:30	Cookout
6:30	Sing-along

Read the schedule to answer the questions.

Activity

1 What will Rod do first? _____

2 Is the nature walk before or after lunch? _____

3 What will Rod do now? _____

4 What time is the nature walk? _____

5 What happens at half past five? _____

6 Will Rod eat lunch after 12:30? _____

7 Will Rod go home at 6:30? _____

MeasureWorks™ • Grade 1 · Time • 12

Make a Schedule

Planning Your Time

Intro & Demo	Activity	Sum It Up
10 min	15 min	10 min

Objective
Make a schedule of after-school activities.

Materials
- none -

Grouping
Whole class, then individuals

Open It Up

Ask: What things do you enjoy doing after school? [Answers will vary.]

List students' responses on the board. Provide an opportunity for students to tell about their after-school activities and what they enjoy about them. You may wish to share information about some of your own activities and interests as well.

Ask: What are some of the things you do every evening? [Sample answers: ride the bus home; brush my teeth before bed; eat dinner; feed the dog; go to bed]

List students' responses on the board.

Demonstrate & Discuss

Say: Let's think about the order of your activities. First, you ride the bus home or someone picks you up. What happens next? [Answers will vary.]

Help students think about the usual order of their after-school activities. Some activities happen immediately after school; some happen after dinner; some are part of a bedtime routine.

Ask: Do some of your activities happen at scheduled times? [Samples: I go to bed at 8:30. My dad picks me up from soccer practice at 6:00.]

Student Activity

Read the directions on the student page aloud to students. Students list their usual after-school activities in order in Column 1 on the student page. If students' schedules vary day by day, have students list their activities for today.

Next, students list an approximate time for each activity. Suggest that students check at home tonight to confirm the accuracy of their schedules, making adjustments as needed.

Informal Assessment

Let classmates compare their schedules to see if they eat dinner or go to bed at similar times. /COMPARE AND CONTRAST/

Have students try to rearrange or change their schedules to find more time for reading. /EVALUATE/

Ask: What is your favorite activity on your schedule? What time is it? [Answers will vary.] /DESCRIBE/

Sum It Up

Say: We made schedules to show our after-school activities. Some of you thought of ways you would like to change your schedule.

Ask: Why is a schedule important? [Samples: It helps me know when to be ready. It helps me think about how I spend my time.] /EVALUATE/

What changes would you like to make to improve your schedule? [Answers will vary.] /EVALUATE/

Science Connection

Seeds have schedules, too! For a seed to grow outside, it must be planted at a certain time of year. Then it takes a set approximate number of days to sprout and to produce flowers or vegetables. Have students check the backs of various seed packets to compare planting dates, sprouting times, and harvest times. Plant some seeds and track their growing schedules.

Name _____

My Schedule

Try This

- What do you usually do after school?
- List 5 to 8 things in order.
- Write a time for each one.

If you don't know the time, estimate.

My Schedule

	What I Do	Time
❶	_____	____ : ____
❷	_____	____ : ____
❸	_____	____ : ____
❹	_____	____ : ____
❺	_____	____ : ____
❻	_____	____ : ____
❼	_____	____ : ____
❽	_____	____ : ____

Am I Late?

Planning Your Time
Wrap Up — 10 min
Project — 45 min

Objective
Summarize learning and use vocabulary related to time.

Materials
- KWL charts
- Student clocks
- Optional: crayons or colored pencils
- Scissors
- Hole punch
- Yarn

Grouping
Whole class, individuals, small groups

Discuss and Sum It Up
- Have students review their KWL charts.
- As a class, discuss what students have learned about time and what they wanted to know.
- Have students complete their charts.

Assessment
See test on page Assessment • 1.

Project

Prepare ahead: Each student will need crayons, yarn, a hole punch, scissors, and a student clock.

Review words related to calendar time, including the days of the week, the months, and the seasons. Next, discuss words related to clock time (hour hand, minute hand, minute, hour, half-hour, digital, and analog).

Ask: Why is it important to be able to tell time? [Sample: to know when to do something or be somewhere]

Tell students that they can make storybooks about a frog who needs some help telling time. Assist students as needed:

1. Color the pictures on the student pages (optional).
2. Cut on the dashed lines.
3. Stack the numbered pages in order. Put the title page on top.
4. Punch holes as indicated, and use yarn to tie the pages together.

Read the story as students follow along in the storybooks they made. Stop as needed to allow students to read the clocks pictured in the story, fill in the times at the bottom of the page, and circle the correct word (late, early, just right).

Read the story again. This time, let volunteers read the dialog of the different animal characters. Have students set their student clocks to show the times in the story.

Suggest that students take their storybooks home to share with family members.

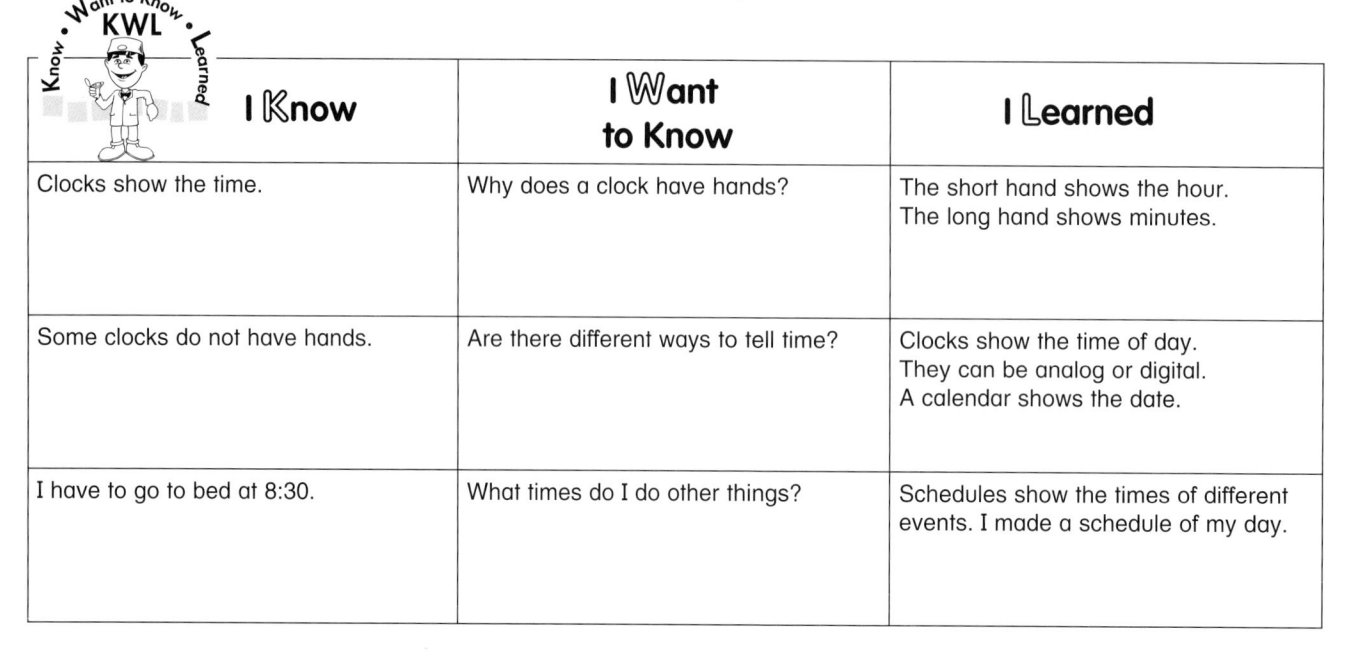

KWL

I Know	I Want to Know	I Learned
Clocks show the time.	Why does a clock have hands?	The short hand shows the hour. The long hand shows minutes.
Some clocks do not have hands.	Are there different ways to tell time?	Clocks show the time of day. They can be analog or digital. A calendar shows the date.
I have to go to bed at 8:30.	What times do I do other things?	Schedules show the times of different events. I made a schedule of my day.

Am I Late?

Name: _____

Frog went to Owl's house.
Here I am, Owl.
You said I could help bake a cake.
You said to come at 1:00.
Am I late?

Frog was supposed to
come at ____ : ____

Frog came at ____ : ____

Frog was ...   early   just right   late

2

No, you are not late.
You are too early.
I am not ready to bake.

Go to Rabbit's house.
Be there at 12:45.
You can help Rabbit buy ice cream.

3

Here I am, Fox.
Rabbit said I could help
blow balloons.
Rabbit said to come at 2:00.
Am I late?

Frog was supposed to
come at ___ : ___ ___.

Frog came at ___ : ___ ___.

Frog was…
late   early   just right

6

I am sorry.
You are too late.
I am done buying ice cream.

*Go to Fox's house.
Be there at 2:00.
You can help Fox blow balloons.*

5

Here I am, Rabbit.
Owl said I could help buy ice cream.
Owl said to come at 12:45.
Am I late?

Frog was supposed to
come at ___ : ___ ___.

Frog came at ___ : ___ ___.

Frog was…
late   early   just right

4

Time • 15   MeasureWorks™ • Grade 1

Surprise! Surprise!
Happy Birthday, Frog!
You are not late.
You are just right!

9

Here I am, Bear.
Fox said to come at 3:00.
Am I late?

Frog was supposed to
come at ___ : ___

Frog came at ___ : ___

Frog was ...
late   early   just right

8

I am sorry.
You are too late.
I am done blowing balloons.

*Go to Bear's house.
Hurry!
Be there at 3:00.*

7

MeasureWorks™ • Grade 1   Time • 16

Answer Key for Time Unit

Time • 1
Clock words: o'clock, afternoon, hour, late, night, minute, early, later, soon, now, morning, night, long hand, short hand
Calendar words: tomorrow, now, yesterday, month, day, week, year

Time • 5
1. 12 o'clock
2. 5 o'clock
3. 8 o'clock
4. 10 o'clock
5. 1 o'clock
6. 4 o'clock

Time • 6
1. 3 o'clock; 4 o'clock; after; before
2. 9 o'clock; 10 o'clock; before; after
3. 11 o'clock; 12 o'clock; after; before
4. 7 o'clock; 8 o'clock; before; after

Time • 7
1. 3:00; Pat the bat
2. 1:30; Dan the dog
3. 8:30; Fran the frog
4. 5:00; Ron the rat
5. 2:30; Matt the cat

Time • 9
1. thirty; 7
2. zero; 5
3. thirty; 11
4. thirty; 2

Time • 12
1. hike
2. after lunch
3. swim
4. 1:30
5. the cookout
6. no
7. no

Time • 14
2. 1:00; 12:15; early

Time • 15
4. 12:45; 1:00; late
6. 2:00; 2:30; late

Time • 16
8. 3:00; 3:00; just right

MeasureWorks™

Length

Unit Introduction

In this unit, students

☞ Explore concepts of length.

☞ Compare and order items of different lengths.

☞ Measure with nonstandard units, inches, and centimeters.

☞ Estimate length.

☞ Use rulers marked in inches and in centimeters.

Assessment
A unit test in multiple-choice format is provided on page Assessment • 2.

KWL

Use a KWL chart to activate prior knowledge and set learning goals as a class. A reproducible KWL chart is provided on page BLM • 7.

Have students keep the KWL chart in their math folders and add to it as they work through this unit.

Games for Practice and Review
Use the MeasureWorks Game Board to reinforce learning. Game rules begin on page BLM • 13.

Focus on Vocabulary

centimeter (p. T-12)	longer (p. T-1)	nearest inch (p. T-11)	tall (p. T-2)
estimate (p. T-8)	longest (p. T-2)	short (p. T-1)	taller (p. T-2)
inch (p. T-10)	meterstick (p. T-12)	shorter (p. T-1)	tallest (p. T-2)
long (p. T-1)	nearest centimeter (p. T-12)	shortest (p. T-3)	unit of measure (p. T-6)

When you introduce one of the vocabulary words, write it on an index card. Display the index cards on a bulletin board and review them periodically.

Heads Up!

Students may not yet realize that size is relative. By handling and comparing objects of different lengths, students learn that a "short" object may be longer than another short item.

Be sure that students get plenty of practice measuring with hand spans, Link 'N' Learn Links, bear counters, and other manipulatives. These early experiences provide a foundation for understanding measurement in inches and centimeters.

Book Nook

There Was an Old Lady Who Swallowed a Fly

by Pam Adams

2000: Child's Play International, Ltd.

This is the old favorite about an old lady who swallows animals of ever-increasing size. Have fun reading the tale and comparing animal lengths!

Length

Manipulatives

Pages	Learning Goals	Inchworms	Inchworms/ Centibugs rulers	Link 'N' Learn Links	Meterstick	PopCubes
T-1–1	Use length vocabulary. Compare the length of two items.					✔
T-2–2	Compare lengths and heights of various objects.			✔		✔
T-3–3	Compare the length of paper shoeprints.					
T-4–4	Compare lengths of various objects. Use length vocabulary.			✔		✔
T-5–5	Measure in uniform nonstandard units.			✔		
T-6–6	Measure with two different nonstandard units. Understand that units of different sizes produce different measurements.			✔		✔
T-7–7	Measure the length of a path using nonstandard units.					
T-8–8	Estimate and measure lengths with nonstandard units.	✔				
T-9–9	Estimate lengths with Link 'N' Learn Links. Measure to check estimates.			✔		
T-10–10	Measure to the nearest inch.	✔				
T-11–11	Measure to the nearest inch. Use a ruler to measure.	✔	✔			
T-12–12	Measure to the nearest centimeter.		✔		✔	
T-13–13	Estimate in centimeters. Measure in centimeters.		✔	✔		✔

MeasureWorks™ • Grade 1

Get to Know Length

Planning Your Time
Intro & Demo	Activity	Sum It Up
10 min	15 min	5 min

Objective
Use length vocabulary.
Compare the length of two items.

Materials
- PopCubes
- Ball of yarn
- Scissors
- Pencils (assorted lengths)

Grouping
Whole class, then individuals

Open It Up
Show a ball of yarn. Have a volunteer tell you where to cut to get a **long** piece of yarn. Stretch out the piece of yarn on the floor. Ask students to name items that are **shorter** and **longer** than the piece of yarn.

Next, have a volunteer tell you where to cut to get a **short** piece of yarn. To show that length is relative, cut an even shorter piece of yarn and compare the two. Help students understand that a "short" item can be longer than another short item.

Demonstrate & Discuss
Distribute a handful of PopCubes to each student. Have each student use several PopCubes to make a train of any length. Have two volunteers with different-length trains show their trains to the class, holding them together to compare length.

Ask: Are the trains the same length? [no] Which train is longer? Which is shorter?

Have each student compare train lengths with another classmate. Ask students to determine which train is shorter, which train is longer, or if both trains are the same length.

Student Activity
Prepare ahead: Each student will need a handful of PopCubes.

Read the directions on the student page aloud to students. Students begin by tracing around their PopCube trains on the student page. You may wish to have students work in pairs, with one partner holding the train as the other traces around it.

Then students work individually to locate and draw an item shorter than the train and an item longer than the train.

Informal Assessment
As students work, encourage them to use the terms *shorter* and *longer*.

Ask: Which object is longer (shorter)? How do you know? [Sample: My train is longer. I compared the objects side by side.] /COMPARE AND CONTRAST/

Sum It Up
Say: Today we talked about length and found objects that are shorter and longer than our PopCube trains.

Ask: How can you tell if something is shorter or longer than another object? [I can compare the two side by side.] /SUMMARIZE/

What did you find that was shorter than your train? What was longer? [Answers will vary.] /DESCRIBE/

Extension
Tell students to begin making a PopCube train when you say *Go*, and stop when you say *Stop*. Make your own train as they work. After you say stop, hold up your train.

Ask: Who has a train that is shorter?

Have a volunteer place his or her train next to yours.

Ask: Who has a train that is longer?

Repeat until all the trains are displayed in order.

Name _____

PopCube Trains

Try This

- Make a train using 🧊.
- Trace around your 🧊 train to show how long it is.
- Find something shorter than your train. Draw it.
- Find something longer than your train. Draw it.

Length means how long.

❶ My train

❷ Shorter than my train

❸ Longer than my train

MeasureWorks™ • Grade 1

Length • 1

Understand Long, Longer, Longest

Planning Your Time
Intro & Demo	Activity	Sum It Up
15 min	15 min	5 min

Objective
Compare lengths and heights of various objects.

Materials
- Link 'N' Learn Links
- PopCubes
- Tools that illustrate length (paint roller, broom, hammer, comb, ice-cream scoop, etc.)
- Tools that illustrate height (briefcase, cooking pot, mop bucket, laundry basket, etc.)
- Tools for teachers (dictionary, planner, etc.)
- Crayons

Grouping
Whole class, then individuals

Open It Up
Display the tools that illustrate length on the floor and the tools that illustrate height on a table. Hold up one of the tools, such as a cooking pot.

Ask: Who might use this tool, and how would they use it? [Sample: A cook might use the pot to boil spaghetti.]

Show and discuss the remaining tools. Compare the length (and height) of several tools.

Demonstrate & Discuss
Show students the tools for teachers (new stick of chalk, pen, long pencil, dictionary, teacher edition of a book, planner).

Say: Here are some of the tools teachers use.

Set the dictionary, teacher edition, and planner on the chalkboard ledge. Ask a volunteer to arrange these items to show **tall, taller, tallest.**

Have a volunteer arrange the remaining tools to show long, longer, and **longest.**

Student Activity
Prepare ahead: Each student will need a handful of Link 'N' Learn Links and a handful of PopCubes.

Read the directions on the student page aloud to students. Students work individually to compare and draw chains of Link 'N' Learn Links and towers of PopCubes. They make one chain that is long, one that is longer, and one that is longest. Then they draw all three chains in the appropriate boxes. They repeat for three PopCube towers.

To help students remember the meanings of tall and long, remind them how they compared the tools on the floor and the table.

Informal Assessment
As students work, help them use math language to describe what they are doing and thinking.

Ask: What chain do you think is longer than this one? [Answers will vary.] /PREDICT/

How could you check to see if you are right? [Sample: You could put the two chains side by side to compare their lengths.] /DESCRIBE/

Sum It Up
Say: Today we compared the lengths and heights of chains and towers. Show me the chains you drew to show long, longer, longest. /SUMMARIZE/

Ask: Which of your towers is tall? taller? tallest? [Have students display their towers.] /SUMMARIZE/

Extension
Trace the outline of each student as he or she lies on a sheet of butcher paper. Have the students color and cut out the "paper people."

Display the paper people along a wall of your classroom, three at a time. Ask students to help you arrange them to show tall, taller, and tallest.

Name _____

Long and Tall

Measure Works

Try This

- Make a long chain of ⚭.
- Make a longer one.
- Make a chain that is longest.
- Draw all three chains in the correct boxes below.
- Make a 🧊 tower that is tall.
- Make a taller one.
- Make a tower that is tallest.
- Draw all three towers in the correct boxes below.

Tall measures up and down. Long measures side-to-side.

① Show long, longer, and longest.

This chain is long.

This chain is longer.

This chain is longest.

② Show tall, taller, and tallest.

This tower is tall. | This tower is taller. | This tower is tallest.

MeasureWorks™ • Grade 1   Length • 2

Understand Short, Shorter, Shortest

Planning Your Time

Intro & Demo	Activity	Sum It Up
15 min	15 min	5 min

Objective

Compare the length of paper shoeprints.

Materials

- Construction paper, assorted colors
- Scissors
- Crayons

Grouping

Whole class, then pairs

Open It Up

Share a book about animal tracks. (See the recommendation in the Science Connection below or use books and encyclopedias from your school library.)

Compare different animal tracks. Find and discuss three tracks that illustrate short, shorter, and **shortest.**

Let students tell about times they have made or found tracks in the snow, in the mud, on the beach, or on the floor.

Demonstrate & Discuss

Ask two volunteers to help you demonstrate how to make paper shoeprints. Sit on a chair with your feet resting on sheets of construction paper. Let the volunteers trace around your shoes. Cut out the shoeprints.

Hold the paper shoeprints side by side so that students can compare their lengths. The shoeprints will probably be about the same size. Because of small tracing differences, one may be shorter than the other.

Tell students that they will be making paper shoeprints. They will compare the shoeprints to find which are short, shorter, and shortest.

Student Activity

Prepare ahead: Each pair will need two sheets of colored construction paper, a pair of scissors, and crayons.

Read the directions on the student page aloud to students. Students work in pairs. They place both feet on a sheet of construction paper, trace around both shoes, and cut out the shoeprints. Repeat with the other sheet of paper so each student has four shoeprints. They write their names on the back of the prints. Collect all of the shoeprints. Let each pair of students select three shoeprints that are different colors and different lengths. (There will be some leftover shoeprints.) Each pair arranges the shoeprints to show short, shorter, and shortest. Students record information about their prints by correctly coloring the shoeprints on the student page.

Informal Assessment

As students work, encourage them to use the vocabulary: *short*, *shorter*, and *shortest*.

Ask: How many of your shoeprints are shorter than this one? [Sample: One is shorter.] /COMPARE AND CONTRAST/

Sum It Up

Say: Today we compared shoeprints and arranged them to show short, shorter, and shortest.

To review, hold up three shoeprints or other items. Ask a volunteer to compare the items and point out short, shorter, and shortest.

Science Connection

Look at various animal tracks in a tracking guide. An excellent guide with life-size tracks is *Mammal Tracks: Life-Size Tracking Guide* by Lynn Levine and Martha Mitchell (Heartwood Press, 2001).

Have each student draw a favorite animal from the book and trace its tracks. As students present their work, talk about tracks that show short, shorter, and shortest.

Name _____

Short, Shorter, and Shortest

Try This

- Choose three shoeprints.
- Put them in order: short, shorter, and shortest.
- Color the shoeprints below to show what you found.

Short can measure up and down or side-to-side.

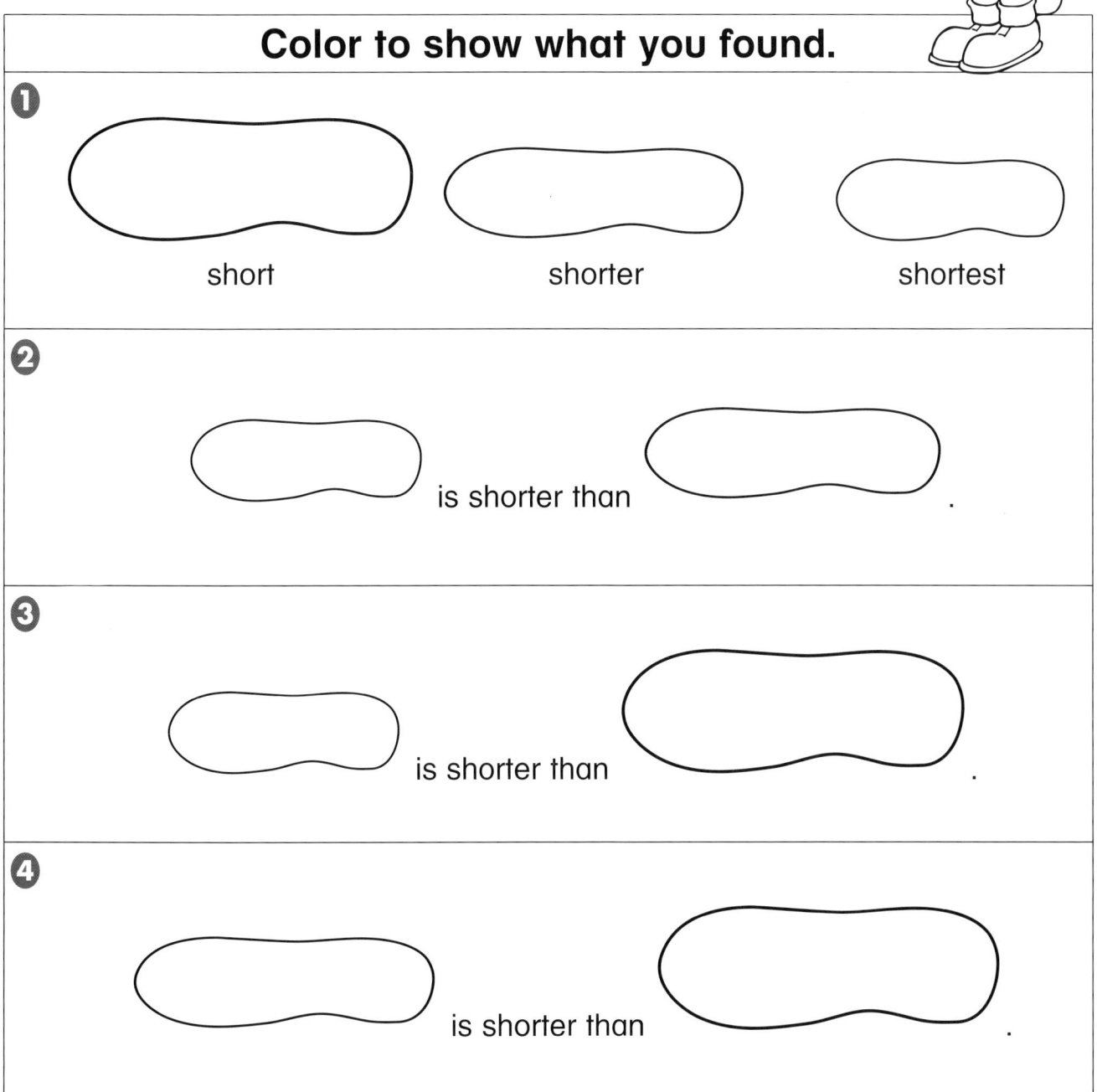

Color to show what you found.

1

short   shorter   shortest

2

_____ is shorter than _____ .

3

_____ is shorter than _____ .

4

_____ is shorter than _____ .

MeasureWorks™ • Grade 1   Length • 3

Compare Length

Planning Your Time
Intro & Demo	Activity	Sum It Up
10 min	15 min	5 min

Objective
Compare lengths of various objects. Use length vocabulary.

Materials
- PopCubes
- Link 'N' Learn Links

Grouping
Whole class, then pairs

Open It Up

Introduce the lesson with a game of "I Spy," incorporating terms for comparing length and height. Begin, for example, with a description of the doorway.

Say: I spy something tall. It is taller than a window. It is the tallest opening in the room.

Provide additional hints if needed until a student correctly guesses the answer. [doorway or door]

Continue by describing something short and something long.

Demonstrate & Discuss

Ask: What are some of the words I used to describe length and height? [Samples: short, shorter, shortest, long, longer, longest, tall, taller, tallest]

As students respond, list the terms on the board.

Ask: Can you find something in the classroom that is short? [Answers will vary.]

Ask volunteers to find something that is shorter and something that is shortest. Draw them on the board.

Repeat for long, longest, and longest.

Explain that the class will have a scavenger hunt.

Ask: If a clue in the scavenger hunt asks for something short, what short items could you find in the classroom? [Samples: PopCube, eraser, paper clip, crayon]

Student Activity

Read the directions on the student page aloud to students. Explain that in a scavenger hunt, everyone gets a list of clues. The players find an item for each clue. The clues in this scavenger hunt will be words that describe length or height.

Explain that students do not need to handle the items they find unless they need to put them side by side to compare lengths.

Provide PopCubes and Link 'N' Learn Links for students to use in addition to classroom objects.

Students work in pairs to find the items and then draw the objects on their individual student pages.

Informal Assessment

As students work, encourage them to use terms for comparing length or height.

Say: Compare the length of these three items. [This one is short (long, tall); this one is shorter (longer, taller); this one is shortest (longest, tallest).] /DESCRIBE/

Sum It Up

Say: Today we found and compared the lengths and heights of different things.

Say: Tell me about the things you found to show short, shorter, shortest. [Sample: A pair of scissors is short, a crayon is shorter, and a paper clip is shortest.] /DESCRIBE/

Then ask what long (longer, longest) and tall (taller, tallest) items students found.

Bulletin Board Idea

Post vocabulary words that describe length and height on the board (or use the list you made in Demonstrate & Discuss).

Ask a volunteer to come to the board, point to one of the words, and use it in a sentence. Repeat for the other words.

Name _____

Compare Length

Try This

- Find three things that show short, shorter, and shortest. Draw them in the correct box below.
- Find three things that show long, longer, and longest. Draw them in the correct box below.
- Find three things that show tall, taller, and tallest. Draw them in the correct box below.

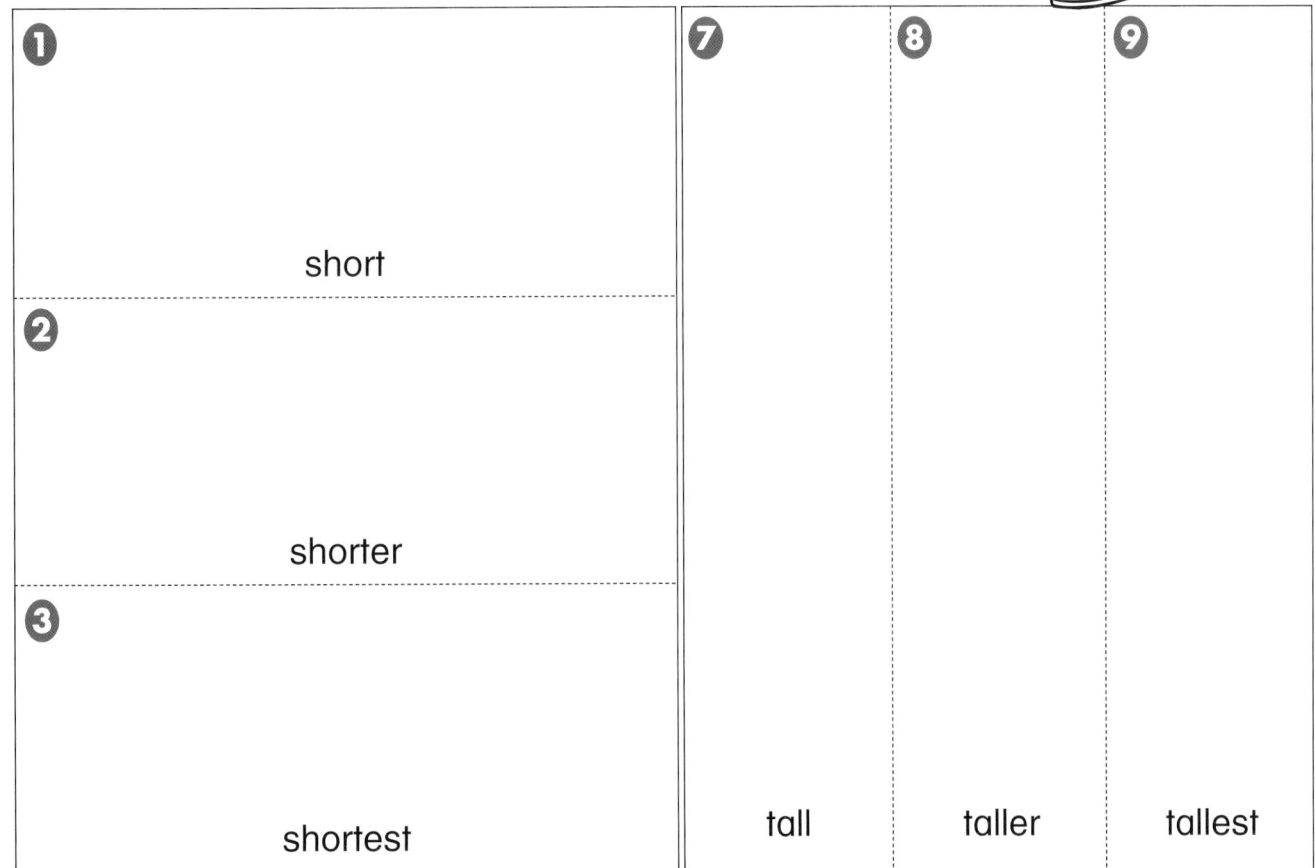

Complete one box at a time.

❶

short

❷

shorter

❸

shortest

❼ **❽** **❾**

tall | taller | tallest

❹

long

❺

longer

❻

longest

MeasureWorks™ • Grade 1 — Length • 4

Measure with Link 'N' Learn Links

Planning Your Time
Intro & Demo: 5 min
Activity: 20 min
Sum It Up: 5 min

Objective
Measure in uniform nonstandard units.

Materials
- Link 'N' Learn Links
- Modeling clay (4 colors per group, varying amounts from marble to golf-ball size)
- Crayons

Grouping
Whole class, then groups of four

Open It Up

Ask: How many of you have ever seen a snake?

Let volunteers tell about the snakes they have seen and the length of the snakes. Tell students that they will be making and measuring modeling-clay snakes.

Demonstrate & Discuss

Place a ball of modeling clay on a desk or table. Demonstrate how to use the palms of your hands to roll the ball into the shape of a snake.

Show some Link 'N' Learn Links (unlinked) to the class.

Ask: How can we find how many links long the snake is? [Sample: Put the links end to end alongside the snake.]

Have students count the links as you put them in place. It may not be possible to make the row of links exactly the same length as the snake. Explain that the measurement is not exact.

Student Activity

Prepare ahead: Each group member will need crayons, a handful of Link 'N' Learn Links, and modeling clay of a different color. The amounts of modeling clay should vary from marble to golf-ball size.

Read the directions on the student page aloud to students. Students work in groups of four.

Each student makes a snake from the modeling clay. Some snakes will be longer than others.

Then students within each group work together to measure the group's snakes. Finally, they record their findings on their student pages.

Informal Assessment

As students work, encourage them to use math language to describe what they are doing.

Ask: How long is this snake? [Sample: It is 6 links long.] /DESCRIBE/

Do you think any of your other snakes will be 6 links long? Why or why not? [Sample: No. The other snakes will be much longer or shorter.] /PREDICT/

Sum It Up

Say: Today we learned how to measure with Link 'N' Learn Links.

Ask: How can you measure an object with Link 'N' Learn Links? [Sample: Place a row of links alongside the object and then count the number of links.] /SUMMARIZE/

Science Connection
Explain that a real snake is covered with dry scales. The snake is constantly growing a new layer of scales. When the outer layer of scaly skin becomes worn, the snake molts by crawling out of its old skin. If possible, show students a snake skin. Alternatively, display a life-sized drawing of a snake. Help students use Link 'N' Learn Links to measure the snake skin or drawing.

Name _____

Clay Snakes

MeasureWorks

Try This

- Make a snake from clay.
- Use 🔗 to measure your snake.
  Measure the other snakes in your group, too.
- Write all the measurements below.

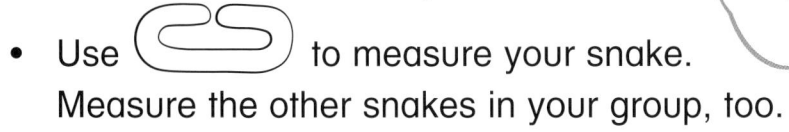

Be sure to let everyone in your group help measure.

Color the snake. Write how long each snake is.

1 The ~~~~ is _____ 🔗 long.

2 The ~~~~ is _____ 🔗 long.

3 The ~~~~ is _____ 🔗 long.

4 The ~~~~ is _____ 🔗 long.

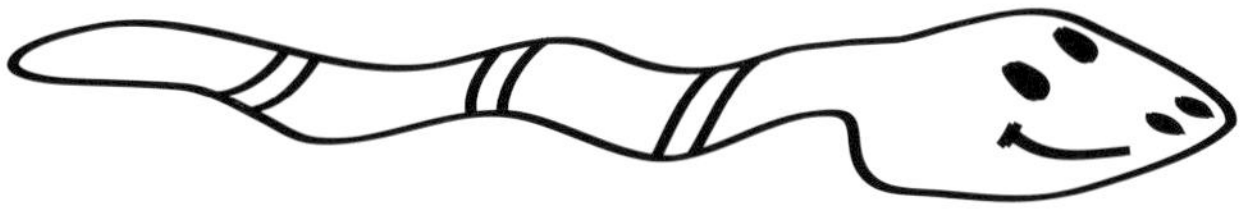

5 Now make a snake 5 🔗 long. Trace it here.

MeasureWorks™ • Grade 1 Length • 5

Explore Units of Measure

Planning Your Time
Intro & Demo: 5 min
Activity: 25 min
Sum It Up: 5 min

Objective

Measure with two different nonstandard units. Understand that units of different sizes produce different measurements.

Materials

- Link 'N' Learn Links
- PopCubes

Grouping

Whole class, then pairs

Open It Up

Remind the class that they measured modeling-clay snakes with Link 'N' Learn Links.

Say: Today we are going to measure the length of the table.

Distribute several Link 'N' Learn Links to each student.

Ask: Would links be a good **unit of measure**? [Samples: Yes, links are all the same size. No, links are short, so it would take a long time to lay them out.]

Have students hold up their Link 'N' Learn Links and guess how many would fit across the table.

Demonstrate & Discuss

Demonstrate how to measure the width of a table. Use a Link 'N' Learn Link as the unit of measure. Have the students count the number of links aloud as you lay them across the table. You are likely to end with a partial link length.

Ask: Do you think we should count the partial link length? [Elicit the response to count the partial length if it is close to a full length or at least half the link length.]

Explain that a hand span is the distance between the tips of the extended thumb and little finger.

Ask: If I measure the table with my hand span, do you think we will get a bigger or smaller number than when we measured with links? [smaller]

Demonstrate how to measure the table with your hand span. Use a finger to temporarily mark the end of the span as you work your way across the table. Have students count aloud.

Student Activity

Prepare ahead: Each pair will need a collection of either Link 'N' Learn Links or PopCubes.

Read the directions on the student page aloud to students. Remind them what a hand span is. Have students stretch out their fingers to show their hand spans.

Students work in pairs to measure the length of a desk, the length of a book, and the length of another item of their choice. Students measure with hand spans and then with either PopCubes or Link 'N' Learn Links. Each student completes the student page by filling in the blanks and writing the measurements in the boxes.

Informal Assessment

Ask: What unit of measure are you using? [Answers will vary.] /COMPREHENSION/
Do you think every student will get the same answer when they measure with hand spans? Why or why not? [Sample: No, because our hand spans are different sizes.] /INFER/

Sum It Up

Say: Today you measured the length of a desk, a book, and some other objects with two different units of measure.

Ask: Did you get a bigger number when you measured with hand spans or with links or PopCubes? Why? [Sample: I got a bigger number when I measured with links because it took a lot more links than hand spans to reach across the room.] /COMPARE AND CONTRAST/

Literature Connection

How Big Is a Foot? by Rolf Myller (Glenview, IL, Scott Foresman Publisher, 1991)

Read this amusing tale about building a bed for the queen. Discuss the problems that characters in the book encounter when they use nonstandard units of measure.

Length • T-6

MeasureWorks™ • Grade 1

Name _____

Ways to Measure

Try This

- Measure the items below.
- First measure in hand spans.
- Then measure with 🧊 or 🔗.

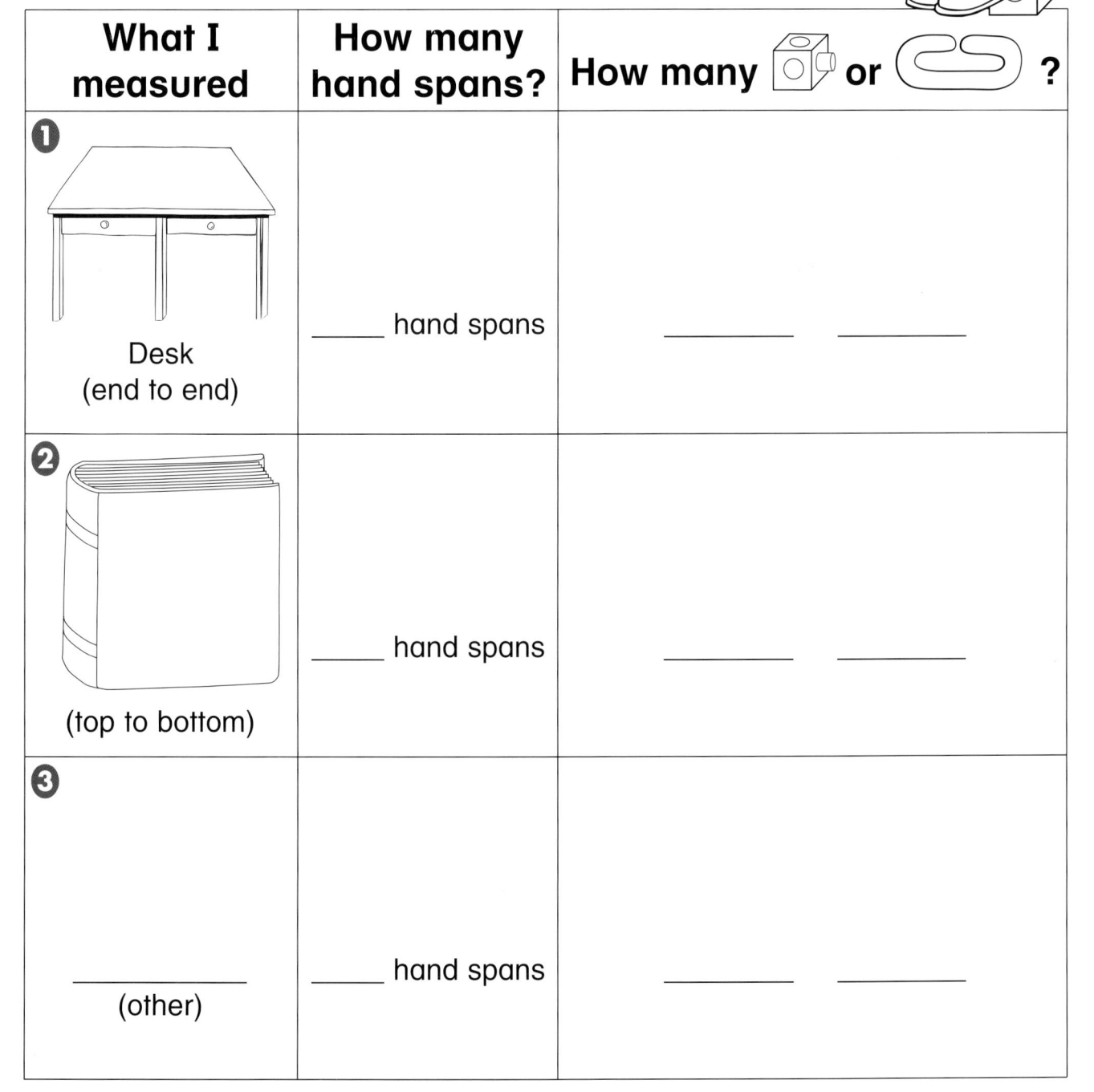

Put the 🧊 or 🔗 end-to-end in a straight line.

What I measured	How many hand spans?	How many 🧊 or 🔗 ?
❶ Desk (end to end)	____ hand spans	_____ _____
❷ (top to bottom)	____ hand spans	_____ _____
❸ _____ (other)	____ hand spans	_____ _____

MeasureWorks™ • Grade 1   Length • 6

Measure with Steps

Planning Your Time

Intro & Demo	Activity	Sum It Up
10 min	15 min	5 min

Objective
Measure the length of a path using nonstandard units.

Materials
- Large items to walk around (see Open It Up)

Grouping
Whole class, then individuals

Open It Up

Students will need some large items to walk around. The items can be in the classroom (rugs, tables, desks, chairs, bookcases), outside (trees, shrubs, playground equipment, shapes marked in chalk on the sidewalk or blacktop), or in the gymnasium (cardboard boxes, sheets of poster board, basketball court markings, shapes marked in masking tape on the floor).

Take the students to one of these areas and point out the different items students can walk around. Ask students to tell you different ways they can measure how far it is around each object.

Demonstrate & Discuss

Demonstrate how to take tiny heel-to-toe steps. Let students practice.

Next, show students how to use heel-to-toe steps to measure the path around an object. Begin at the corner of a chair or other object. Circle the object as the students count your heel-to-toe steps aloud.

Ask: Why is it better to measure with heel-to-toe steps instead of regular steps? [Not all my regular steps are the same length.]

Student Activity

Prepare ahead: Make sure the large items to walk around from Open It Up are still available.

Read the directions on the student page aloud to students. Students use heel-to-toe steps to measure the paths around four different items.

Before students begin, remind them that they need to notice and remember exactly where on the path they begin measuring. Usually, the student can begin at a corner of the item. However, if the item is circular or curved, the student may need to mark the starting point by placing a pencil on the ground.

Informal Assessment

As students work, help them use math language to describe what they are doing and thinking.

Ask: How does this path compare in length to the other paths you have measured? [Sample: It is thirty steps long; my other paths were much shorter.] /COMPARE AND CONTRAST/

If you used regular steps instead of heel-to-toe steps, how would the number change? [It would be smaller because I could walk the length in fewer steps.] /INFER/

Will everyone measure the same number of steps? [No, some feet are longer or shorter than others.] /PREDICT/

Sum It Up

Say: Today we used heel-to-toe steps to measure the paths around different items.

Ask: How does measuring the path around an item differ from measuring the length of an item? [Samples: You start and stop at the same place. You measure around all sides, not just one.] /COMPARE AND CONTRAST/

Bulletin Board Idea
How Long is This Path?

Draw a large rectangular path on the bulletin board. Have students use various tools (pencils, crayons, hand spans, index cards, etc.) to measure its length.

Post students' responses.

Name _____

Step by Step

Try This

- Use heel-to-toe steps to measure a path around something. How many steps long is the path?
- Tell about it in a box below.
- Measure and tell about three other paths.

When you measure around something, mark your starting place!

Draw the object. Write how many steps.

❶ A path around

is _____ steps long.

❷ A path around

is _____ steps long.

❸ A path around

is _____ steps long.

❹ A path around

is _____ steps long.

MeasureWorks™ • Grade 1   Length • 7

Estimate with Nonstandard Units

Planning Your Time
Intro & Demo	Activity	Sum It Up
5 min	20 min	5 min

Objective
Estimate and measure lengths with nonstandard units.

Materials
- Inchworms
- Grocery sack containing 4 items to measure (1 sack per group of four)
- Crayons
- Hole punch

Grouping
Whole class, then groups of four

Open It Up
Write **estimate** on the board. Explain that an estimate is a best guess.

Show the class a hole punch and an Inchworm.

Ask: How many worms long do you think this hole punch is? Estimate. [Answers will vary.]

Write students' estimates on the board.

Demonstrate & Discuss
Ask: How can we tell if any of these estimates are correct? [Measure the length of the hole punch with Inchworms.]

Place the hole punch on a table or desk. Have students count aloud as you measure by placing Inchworms, end-to-end, in a row next to the hole punch.

Ask students to compare the actual measurement to the estimates on the board. Have volunteers circle the estimates that match or almost match the actual measurement.

Student Activity
Prepare ahead: Each group will need a handful of Inchworms and a grocery sack containing four items to measure, for example: feather, spoon, candle, and piece of yarn.

Students work in groups of four. Read the directions on the student page aloud to students. Explain that the group members should remove one item at a time from the sack. Each student in the group draws the item and estimates its length. Then students in the group work together to measure the item with Inchworms.

Informal Assessment
As students work, help them use math language to talk about what they are doing and thinking.

Ask: Are you estimating or measuring right now? [Sample: I am estimating.] /COMPREHENSION/

How does knowing the feather's length in Inchworms help you estimate the yarn's length? [Sample: The number of Inchworms will be greater for the yarn because the yarn is longer than the feather.] /INFER/

Sum It Up
Say: Today we estimated the length of different objects. We measured to check our estimates.

Say: Tell me about one of your best estimates. [Sample: I estimated that the feather was 7 Inchworms long. I measured and found that the estimate was right.] /SUMMARIZE/

Literature Connection
There Was an Old Lady Who Swallowed a Fly by Pam Adams (New York: Child's Play International, Ltd., 2000)

After reading the book aloud, have students estimate the length of each animal in Inchworms. Have students measure to check the estimates.

Name _____

Grab Bag

Try This

- Take something from your sack. Draw it in the first box.

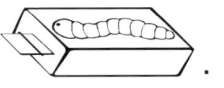

- Estimate its length in 🪱.
- Write your estimate.

- Measure its length. Use 🪱.
- Write the measurement.

- Repeat with the other items in your sack.

Picture one 🪱. Then imagine how many 🪱 you need to measure the objects.

Drawing	Estimated Length	Actual Length
❶	_____ 🪱	_____ 🪱
❷	_____ 🪱	_____ 🪱
❸	_____ 🪱	_____ 🪱
❹	_____ 🪱	_____ 🪱

MeasureWorks™ • Grade 1   Length • 8

Estimate and Check

Planning Your Time

Intro & Demo	Activity	Sum It Up
15 min	15 min	5 min

Objective

Estimate lengths with Link 'N' Learn Links. Measure to check estimates.

Materials

- Link 'N' Learn Links
- Red, yellow, blue, and green paper strips (1 in. by 18 in., one of each color per student)
- Scissors

Grouping

Whole class, then individuals

Open It Up

Play an estimation game. Line up three links on the overhead. Flash the overhead on and off. Hold up a pencil.

Ask: Is this pencil longer or shorter than the line of links?

Repeat with other numbers of links and other objects.

Demonstrate & Discuss

Tell students that they will estimate lengths today. Hold up the chalkboard eraser for the class to see.

Say: Imagine a line 9 erasers long.

Ask a volunteer to estimate the line's length and to draw a line on the board.

Ask: Do you agree that this line is 9 erasers long? Ask several of the students who do not agree to draw their own lines on the board.

To find which line is closest to 9 erasers in length, have volunteers help you use the eraser to measure. Explain that practice helps us get better at estimating.

Student Activity

Prepare ahead: Each student will need four 1 in. by 18 in. paper strips, one each of red, yellow, blue, and green. Each student will also need a pair of scissors and a handful of Link 'N' Learn Links.

Read the directions on the student page aloud to students. Students work individually to estimate a length, cut a construction paper strip to that length, and then measure to check the estimate. They use Link 'N' Learn Links as the unit of measure.

Guide students through Exercise 1 on the student page. Then let them complete the remaining exercises on their own. Remind students to throw away their paper scraps immediately after cutting so that they do not confuse their scraps with their other paper strips.

Informal Assessment

As students work, encourage them to use math language to describe what they are doing.

Ask: How can you tell if your estimate is a good one? [Answers will vary. Elicit the response that a good estimate is one in which the estimated length and the measured length match or almost match.] /INFER/

Did your estimates get better as you made more guesses? [Yes, as I practiced, I got better.] /COMPREHENSION/

Sum It Up

Say: Today we cut paper strips to show different estimated lengths. Then we measured the strips to check our estimates.

Ask: What are some tips for estimating accurately? [Samples: Imagine a row of Link 'N' Learn Links next to the paper strip. Use the answers you get when you measure to help you estimate the next length.] /GENERALIZE/

Extension

Play an estimation game: Specify a unit of measure and have each student estimate the length of an object, such as a window. Measure the object so that the students can check their estimates. The student(s) whose estimate is closest can select the next unit of measure and the next object to measure.

Length • T-9

MeasureWorks™ • Grade 1

Estimate and Check

Name _____

Try This

- Read the first activity.
  Estimate the distance.
  Cut your red paper to that length.
- Measure your paper.
  Was your estimate close?
- Try the other activities.

Don't get mixed up. Toss your scraps after you cut your paper strips.

Estimate and Cut.	Measure.
1 Estimate 4 📎. Cut your red paper. Measure. How many 📎? _____ 📎	**2** Estimate 12 📎. Cut your blue paper. Measure. How many 📎? _____ 📎
3 Estimate 7 📎. Cut your yellow paper. Measure. How many 📎? _____ 📎	**4** Estimate 18 📎. Cut your green paper. Measure. How many 📎? _____ 📎

MeasureWorks™ • Grade 1   Length • 9

Measure with Inchworms

Planning Your Time
Intro & Demo	Activity	Sum It Up
10 min	10 min	5 min

Objective
Measure to the nearest inch.

Materials
- Inchworms

Grouping
Whole class, then individuals

Open It Up

Ask: What are some of the units of measure we have been using? [hand spans, Link 'N' Learn Links, heel-to-toe steps, worms]

Say: Can you name some units of measure that you might find on rulers, measuring sticks, or tape measures? [Samples: inch, foot, yard, centimeter, meter]

Tell students that today they will measure in **inches.** However, instead of using a ruler marked in inches, they will be using Inchworms that are each exactly one inch long.

Demonstrate & Discuss

Give each student six Inchworms.

Say: You have already used these to measure objects from a grab bag. Now you know that each worm is one inch long. This is why we call them Inchworms.

Direct students to place their Inchworms end-to-end on their desks.

Have students point to the Inchworms in order and count the number of inches aloud: "1 inch, 2 inches, 3 inches, 4 inches, 5 inches, 6 inches."

Ask: Can you describe how to use Inchworms to measure an object? [Sample: Line up Inchworms alongside the object and count them.]

Student Activity

Prepare ahead: Each student will need six Inchworms.

Read the directions on the student page aloud to students. Students work individually. They use Inchworms to measure the lines on the student page. Make sure students understand that each Inchworm is exactly one inch in length.

Informal Assessment

As students work, encourage them to use math language to talk about what they are doing and thinking.

Ask: What unit of measurement are you using and how long is it? [I'm using an Inchworm; it is one inch long.] /COMPREHENSION/

(Pointing at two lines) Is this line shorter or longer than this other line on the student page? [Answers will vary.] /COMPARE AND CONTRAST/

Since this line is longer than the other one, will the number of inches be greater or less than before? [greater] /INFER/

Sum It Up

Say: Today we used Inchworms to find how many inches long these lines were.

Ask: How is measuring with Inchworms the same as measuring with inches? [Sample: An Inchworm and an inch are the same length.] /COMPARE AND CONTRAST/

Say: Imagine a line one inch long. Draw it on the back of your student page. Check the length by measuring with an Inchworm.

Social Studies Connection
Explain that the use of an inch as a unit of measure began over seven hundred years ago with a law made by King Edward II of England. An inch was the same length as three grains of barley laid end to end. If possible, compare three grains of barley to an Inchworm manipulative.

Name _____

Inchworms

MeasureWorks™

Try This

- Use [inchworm] to measure each line.
- Write how many inches long each line is.

Remember, Each [inchworm] is one inch long.

Measure. Write how many inches.

1

_____

How long is it? _____ inches

2

_____

How long is it? _____ inches

3

_____

How long is it? _____ inches

4

_____

How long is it? _____ inches

MeasureWorks™ • Grade 1   Length • 10

Measure with Rulers

Planning Your Time

Intro & Demo	Activity	Sum It Up
10 min	20 min	5 min

Objective

Measure to the nearest inch.
Use a ruler to measure.

Materials

- Inchworms/Centibugs rulers
- Inchworms
- Strips of paper 5 in., $5\frac{1}{4}$ in., $5\frac{1}{2}$ in., and $5\frac{3}{4}$ in. in length (1 in. wide)
- Everyday classroom objects (1 to 12 in. long)

Grouping

Whole class, then pairs

Open It Up

Distribute Inchworm rulers and Inchworms to students.

Tell students to line up an Inchworm between the 1 and the 2 on the ruler. Point out that an Inchworm is the same length as the space between the 1 and the 2.

Say: The space between each pair of numbers on the ruler is 1 inch long.

Have students point to the 1-inch segments on their Inchworm ruler, one at a time, and count them aloud.

Tell students that today they will use rulers to measure length in inches.

Demonstrate & Discuss

Gather students around to watch the demonstration. Demonstrate how to measure a 5-inch strip of paper by lining up the end of the strip and the end of an Inchworm ruler. **Ask:** How many inches long is the paper? [5 inches]

Now place a $5\frac{3}{4}$-inch strip of paper next to the Inchworm ruler. Tell students that they will be measuring to the **nearest inch.**

Ask: Is this strip of paper closer to 5 inches or 6 inches in length? [6 inches] Explain that the strip of paper is about 6 inches long.

Continue with a strip of paper $5\frac{1}{4}$ inches long (about 5 inches long) and with a strip $5\frac{1}{2}$ inches long (about 6 inches long).

Student Activity

Prepare ahead: Each student will need an Inchworm ruler. On a table, display assorted classroom objects that are 1 to 12 inches long.

Read the directions on the student page aloud to students. Students work in pairs. The students in each pair select an object on the table, take it to their work area, measure to the nearest inch, and record the information on their student pages.

Informal Assessment

As students work, encourage them to use math language to explain what they are doing.

Ask: How can you find how many inches long this object is? [Sample: The end of the object is close to the 3-inch mark on the ruler. The object is about 3 inches long.] /DESCRIBE/

Sum It Up

Say: Today we used rulers to measure the length of objects to the nearest inch.

Ask: What does it mean to measure to the nearest inch? [Sample: When you give the measurement of the object, you use the number of inches that is closest to the actual length.] /SUMMARIZE/

Bulletin Board Idea

LESS THAN 6 INCHES LONG	MORE THAN 6 INCHES LONG

Let students draw pictures or write words to show items that belong on each side.

Name _____

Measure with Rulers

Try This

- Take an object from the table.
- Measure its length to the nearest inch.
- Draw the object and record its length.
- Put the object back. Repeat with a different object.

Remember to measure to the nearest inch.

Draw the object.	Measure to the nearest inch.
1	_____ inches
2	_____ inches
3	_____ inches
4	_____ inches

MeasureWorks™ • Grade 1   Length • 11

Measure in Centimeters

Planning Your Time

Intro & Demo	Activity	Sum It Up
10 min	15 min	5 min

Objective
Measure to the nearest centimeter.

Materials
- Meterstick
- Inchworms/Centibugs ruler
- Drinking straws
- Scissors
- Transparent tape

Grouping
Whole class, then individuals

Open It Up

Show students a **meterstick.** Explain that it is folded to make it easier to store. Demonstrate how to open it up to full length. Let students tell about similar measuring devices they may have seen (folding carpenter rulers, retractable measuring tapes, and so on).

Say: Some rulers and measuring sticks are divided into inches. Others are divided into **centimeters.**
Point out the 1-centimeter segments on the meterstick.

Demonstrate & Discuss

Distribute Inchworms/Centibugs rulers. Point out the centimeter markings. Explain that this tool has two names: the Inchworm ruler or the Centibug ruler, depending on which side is facing up.

Explain that each bug is 1 centimeter long.

Have students show the length of 1 centimeter with their thumb and forefinger. Then have then put their hands next to the rulers to check.

Ask volunteers to use the rulers to measure classroom objects, such as pencils or books.

It is likely that some of the items will not be an even number of centimeters. Tell students they should measure to the **nearest centimeter.**

Ask: If the length of the object is between two different centimeter markings on the meterstick, what should you do? [Sample: Use the number of centimeters that is closer to the actual length of the object.]

Student Activity

Prepare ahead: Each student will need a Centibug ruler, a drinking straw, and a pair of scissors.

Read the directions on the student page aloud to students. Guide students to measure the straw in Exercise 1 on the student page with the Centibug rulers.

Point out the fact that no straws are shown in Exercises 2 and 3. Direct students to cut their straw into two short straws (any lengths). Help the students tape their straws in place.

Using the Centibug ruler, students work individually to measure and record the lengths of their straws to the nearest centimeter.

Informal Assessment

As students work, encourage them to use math language to describe what they are doing.

Ask: How did you know to write 7 centimeters instead of 6 centimeters? [Sample: The straw was between 6 and 7 centimeters in length, but it was closer to 7 centimeters.] /DESCRIBE/
Are centimeters longer or shorter than inches? [shorter] /COMPARE AND CONTRAST/

Sum It Up

Say: Today we measured the lengths of straws to the nearest centimeter with the Centibug rulers.

Ask: Do you think centimeters are a better unit of measure than heel-to-toe steps? Why or why not? [Elicit that centimeters are better because they are always the same length.] /COMPARE AND CONTRAST/

Science Connection

The metric system of measurement was invented by a group of scientists in France. Today it is used by scientists everywhere to measure length, weight, and temperature. Most scientists prefer the metric system over the customary system (the system that uses inches) because it is grounded in base ten and because it is used around the world.

Length • T-12

MeasureWorks™ • Grade 1

Name _____

Centibug Straws

MeasureWorks

Try This

- How long is the straw in the picture?
  Record its length to the nearest centimeter.

- Cut a real straw into two short straws.
  Tape the short straws below.

- How long are your straws?
  Record their lengths to the nearest centimeter.

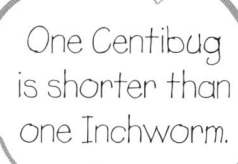

One Centibug is shorter than one Inchworm.

1

| 1 | 2 | 3 | 4 | 5 | 6 | 7 | 8 | 9 | 10 | 11 | 12 | 13 | 14 | 15 | 16 | 17 |

_____ centimeters long

2

(Tape straw here.)

_____ centimeters long

3

(Tape straw here.)

_____ centimeters long

MeasureWorks™ • Grade 1   Length • 12

Estimate in Centimeters

Planning Your Time
Intro & Demo	Activity	Sum It Up
10 min	15 min	5 min

Objective
Estimate in centimeters.
Measure in centimeters.

Materials
- Link 'N' Learn Links
- PopCubes
- Inchworms/Centibugs ruler
- Books

Grouping
Whole class, then pairs

Open It Up

Say: Hold up your thumb and index finger so that the distance between them is about a centimeter.

Observe whether students show an appropriate length.

Have students sitting next to each other measure the length to check the estimate.

Ask: Are any of your fingernails about a centimeter wide or a centimeter long? Measure with your Centibug ruler to check.

Ask volunteers to share the benchmarks they found on their hands.

Demonstrate & Discuss

Make a tower four PopCubes tall and rest one end of a book on top of it to make a ramp. Hold a Link 'N' Learn Link near the top of the book. Release it so it slides down the ramp.

Ask: How far do you think the link slid from the bottom of the ramp to where it is now? Estimate in centimeters. [Answers will vary.]

Have students come to the front of the room to observe how far the link traveled. Encourage them to compare the 1-centimeter lengths on their fingernails with the distance to estimate.

When students return to their seats, tell them to raise their hands when they agree with the estimate.

Ask: Do you think the link traveled about 1 centimeter? 5 centimeters? 7 centimeters? 10 centimeters? 15 centimeters? 20 centimeters? [Answers will vary.]

Have a volunteer measure with a Centibug ruler to check.

Student Activity

Prepare ahead: Each pair will need five PopCubes, a book, a Link 'N' Learn Link, and a Centibug ruler.

Read the directions on the student page aloud to students. Students work in pairs. They make a tower of four PopCubes and place one end of a book on the tower to make a ramp. They let a Link 'N' Learn Link slide down the ramp until it comes to a stop. Then they estimate how far the link traveled from the bottom of the book. They measure with a Centibug ruler to check their estimate. They repeat once more. Then they add a PopCube to the tower and make two more trials.

Informal Assessment

As students work, encourage them to use previous measurements to help them make new estimates.

Ask: How could you use your fingernail to help you estimate? [Sample: I know that my fingernail is about a centimeter wide, so I hold it next to the link and imagine how many times it would fit between the link and the book.] /DESCRIBE/

Sum It Up

Say: Today we learned how to estimate in centimeters and measure to check our estimates.

Ask: Did the link travel farther when you used four PopCubes or when you used five?
[when I used five] /COMPARE AND CONTRAST/
When do you estimate distance in real life?
[Samples: When I play catch, I estimate how far the ball will go; when I eat dinner, I estimate how much I can eat.] /APPLY/

Literature Connection

Frog and Toad Are Friends by Arnold Lobel (Harper Collins Juvenile Books, 1979)

Read the story aloud. On a page with large illustrations, have volunteers measure the height of the characters in centimeters. Discuss how long a bed each character needs, how tall a chair, how high a table, etc. Have students estimate in centimeters.

Length • T-13

MeasureWorks™ • Grade 1

Name _____

Sliding Links

The letters "cm" stand for centimeter.

Try This

- Make a tower of four ⬚.
- Lean a book on the tower to make a ramp.
- Hold a ⌒ at the top of the ramp.
- Let the ⌒ go.
- Estimate how far the ⌒ slid from the bottom of the book. Record.
- Measure to check. Record.
- Repeat.
- Add a ⬚ to the tower.
- Repeat to complete the chart.

How High?	Estimate	Measure
❶ 4 PopCubes	_____ cm	_____ cm
❷ 4 PopCubes	_____ cm	_____ cm
❸ 5 PopCubes	_____ cm	_____ cm
❹ 5 PopCubes	_____ cm	_____ cm

Measurement Man

Planning Your Time
Wrap Up 15 min
Project 60 min

Objective
Reinforce concepts of length.

Materials
- KWL charts
- Crayons
- Scissors
- Glue
- Inchworms/Centibugs rulers
- Link 'N' Learn Links

Grouping
Whole class, then pairs

Discuss and Sum It Up

- Have students review their KWL charts.
- As a class, discuss what students knew about length and what they wanted to know.
- Have students share and record what they have learned.

Assessment
See test on page Assessment • 2.

Project

Prepare ahead: Each pair will need a pair of scissors, glue or paste, an Inchworm ruler, and a handful of Link 'N' Learn Links.

Have students draw Measurement Man's face and color the parts on page Length • 14. Then have students cut on the solid lines, fold on the dashed line, and assemble:

1. Glue the ends together. 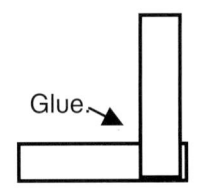
2. Fold one strip and then the other. Keep repeating. 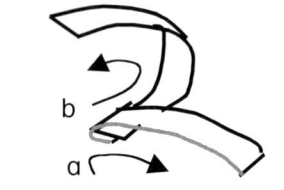
3. Glue the ends together. 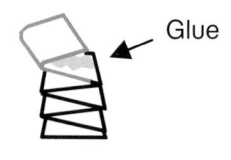
4. Glue on the arms. Then glue on the head and feet.

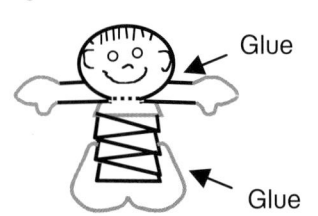

Let students work in pairs to complete page Length • 15. One student in each pair can steady Measurement Man while the other student takes measurements. Students measure to the nearest unit with Inchworm rulers on both the Inchworm and Centibug sides and with Link 'N' Learn Links.

Students work individually to complete the estimates of real-life classroom objects on page Length • 16. Students should use their actual Measurement Men as a unit of measure. Then check the estimates as a class, using everyone's Measurement Men.

KWL

I Know	I Want to Know	I Learned
Things are different lengths.	What things are long, and what things are short?	A pencil is long compared to a crayon. It is short compared to a window.
People measure to tell how long things are.	How can I measure things?	I can measure by using Inchworms, crayons, inches, centimeters, or many other things.
At the clinic, I stood by a ruler to get measured.	Why do people use rulers to measure?	Rulers have units that are always the same length.

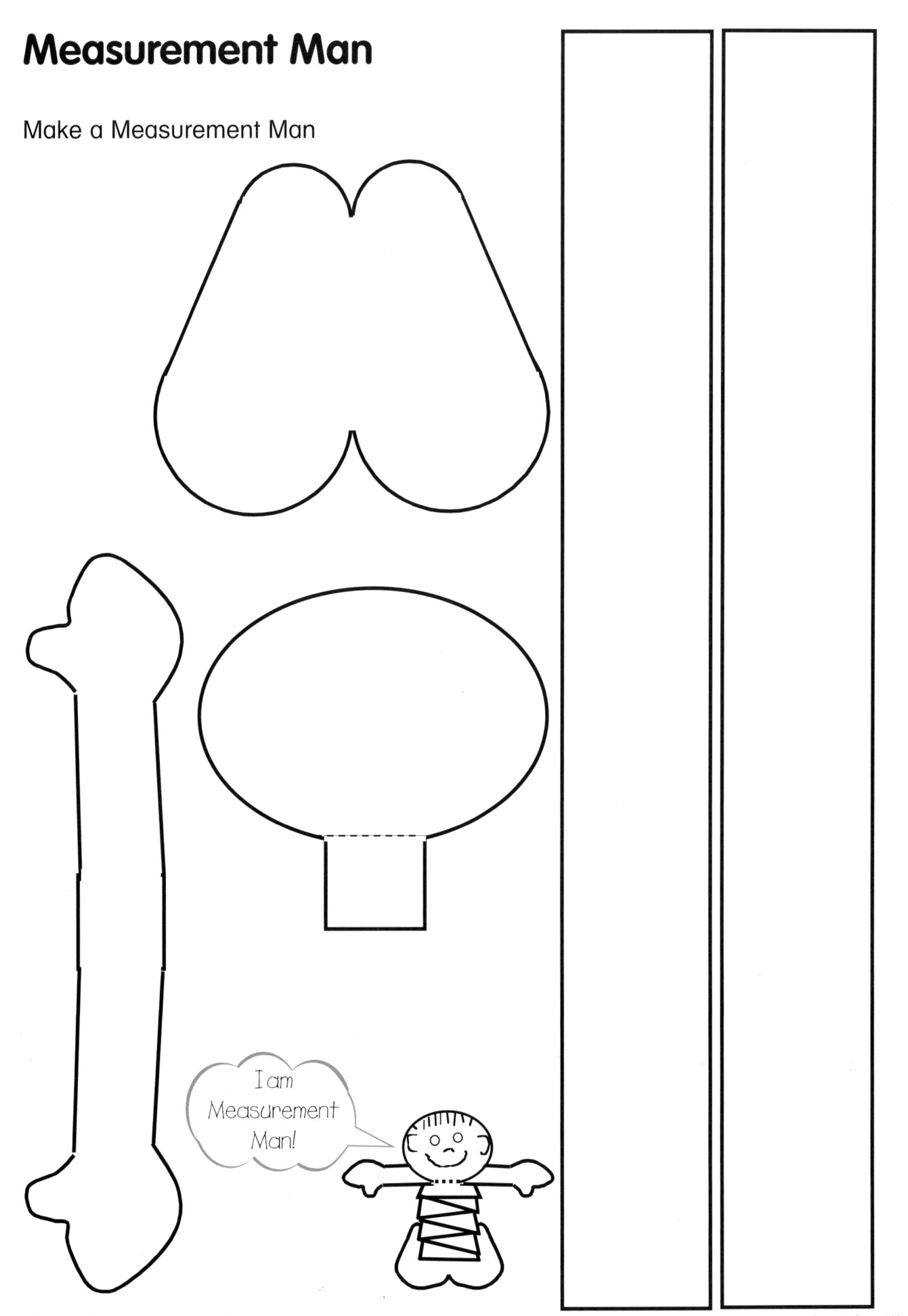

Measurement Man

Make a Measurement Man

I am Measurement Man!

MeasureWorks™ • Grade

Length • 14

Name _____

Measurement Man, continued

Measure the Measurement Man that you made. The pictures show what to measure.

Measurement Man's Feet

How wide: _____ ⌇

_____ inches

Measurement Man's Arm

How long: _____ ⌇

_____ centimeters

Measurement Man's Head

How wide: _____ ⌇

_____ centimeters

Measurement Man

How tall: _____ ⌇

_____ inches

Push Measurement Man down.

How short: _____ ⌇

_____ inches

Pull Measurement Man up.

How tall: _____ ⌇

_____ inches

Length • 15

MeasureWorks™ • Grade 1

Name _____

Measurement Man, continued

Measurement Man Estimation

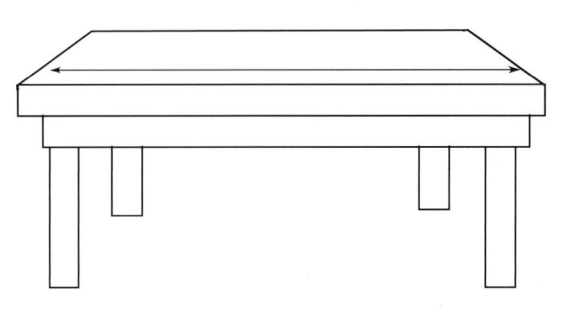

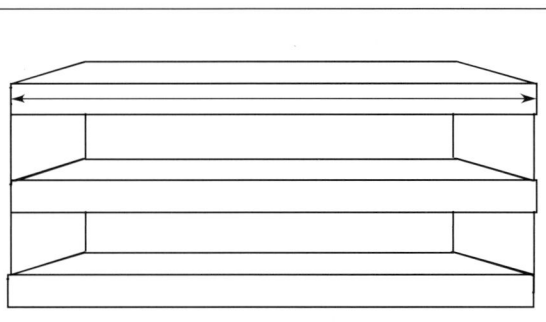

	Estimate: _____ Measurement Men Measure: _____ Measurement Men
	Estimate: _____ Measurement Men Measure: _____ Measurement Men
	Estimate: _____ Measurement Men Measure: _____ Measurement Men
	Estimate: _____ Measurement Men Measure: _____ Measurement Men

MeasureWorks™ • Grade 1   Length • 16

Answer Key for Length Unit

Length • 10
1. 3 inches
2. 5 inches
3. 4 inches
4. 2 inches

Length • 12
1. 15 centimeters long
2. Answers will vary.
3. Answers will vary.

MeasureWorks™
Capacity

Unit Introduction

In this unit, students

☞ Explore concepts of capacity.

☞ Discover that size and capacity are different measures.

☞ Estimate and then compare capacities of containers.

☞ Estimate and then use capacity to order containers.

☞ Use standard units of measure: one cup and one liter.

☞ Identify containers that measure exactly one cup or one liter.

Assessment
A unit test in multiple-choice format is provided on page Assessment • 3.

KWL

Use a KWL chart to activate prior knowledge and set learning goals as a class. A reproducible KWL chart is provided on page BLM • 7.

Have students keep the KWL chart in their math folders and add to it as they work through this unit.

Games for Practice and Review
Use the MeasureWorks Game Board to reinforce learning. Game rules begin on page BLM • 13.

Focus on Vocabulary

capacity (p. T-1)	less (p. T-3)	more (p. T-3)	standard measure (p. T-5)
cup (p. T-6)	less than (p. T-2)	more than (p. T-2)	
least (p. T-3)	liter (p. T-8)	most (p. T-3)	

Post the vocabulary words on a bulletin board. When you or a student uses one of the words in daily context, point to the word, then have students say the word aloud. Alternatively, point to one of the words, say it, have students say it aloud, and then invite a volunteer to use the word in a sentence about capacity. Repeat with other words. Model proper sentence structure as needed.

Heads Up!

Capacity is what a container can hold. Volume is the amount of space an object takes up. Often, students use these terms interchangeably. This unit focuses mostly on capacity.

Students often relate capacity to the size and the shape of the container. They sometimes think that the capacity of a tall, narrow container is more than the capacity of a short, wide container. Filling containers of different sizes and shapes allows students to visualize capacity and discover that taller does not always mean more.

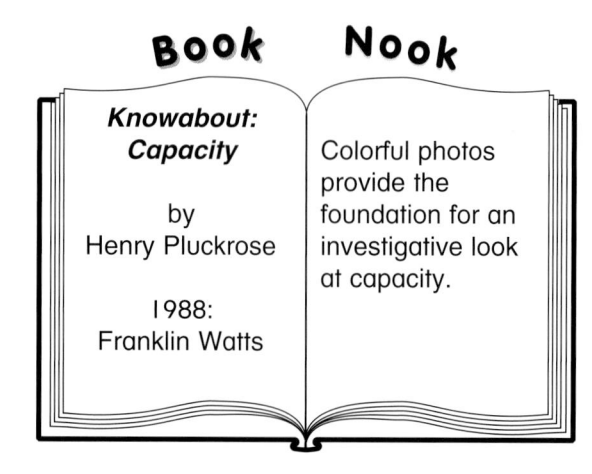

Book Nook

Knowabout: Capacity

by
Henry Pluckrose

1988:
Franklin Watts

Colorful photos provide the foundation for an investigative look at capacity.

Capacity

Manipulatives

Pages	Learning Goals	Bear counters	Measuring cups	PopCubes
T-1–1	Use nonstandard units to estimate capacity. Compare estimates to outcomes.	✔		✔
T-2–2	Use *more than* and *less than* to compare capacity.			
T-3–3	Compare capacity. Use *more*, *most*, *less*, and *least* to describe capacity.			✔
T-4–4	Estimate capacity in nonstandard units. Measure to compare capacity.			
T-5–5	Understand the need for standard units of measure.	✔	✔	
T-6–6	Explore math tools. Use a 1-cup measuring cup to measure capacity of containers.		✔	
T-7–7	Use manipulatives to estimate capacity. Compare estimates to actual measures.		✔	
T-8–8	Explore capacity measurement tools. Use a 1-liter measuring tool.			

MeasureWorks™ • Grade 1   Capacity • T-1B

Get to Know Capacity

Planning Your Time

Intro & Demo	Activity	Sum It Up
10 min	10 min	5 min

Objective
Use nonstandard units to estimate capacity. Compare estimates to outcomes.

Materials
- PopCubes
- Bear counters
- A variety of containers (transparent storage containers, tin cans, yogurt containers, etc.)

Grouping
Whole class, then groups of three

Open It Up

Display a variety of containers. Point out that each container is a different size and shape. Explain that each container can hold a different number of PopCubes.

Fill each container with PopCubes, then **ask:** Do each of these containers hold the same number of PopCubes? [no] How do you know? [Sample: We can count PopCubes.]

Discuss how many PopCubes each container holds.

Empty the containers, then repeat with bear counters.

Explain that we use the word **capacity** to tell how much a container can hold.

Demonstrate & Discuss

Make a two-column chart on the board with the heads: *Estimate* and *Count.*

Explain that an estimate is a reasonable guess.

Give a volunteer a container, then **ask:** How many PopCubes will fit in your container? Record the response in the *Estimate* column.

Have the class count aloud as the volunteer drops PopCubes in the container one at a time. When the container is full, record the number in the count column.

Repeat with other volunteers, containers, and objects. Guide students to see that the count tells how much the container holds.

Capacity • T-1

Student Activity

Prepare ahead: Provide a collection of containers for the class to use. Each group of three will need a set of PopCubes and a set of bear counters.

Read the directions on the student page aloud to students. Each student in a group selects a container and draws it in the *Draw* column. Each student estimates how many PopCubes his or her container will hold and records the estimate. A student drops PopCubes one at a time in his or her container while all group members count aloud by 1s. When the container is full, the student records the actual count. Repeat with the second and third group member.

Students empty their containers and repeat the activity with bear counters.

Informal Assessment

As students work, notice if their estimates show an understanding that filling a container with larger objects (PopCubes) results in a smaller number than filling a container with smaller objects (bears).

Ask: Why do you need fewer PopCubes than bear counters to fill your container? [Sample: The PopCubes take up more space.] /DRAW CONCLUSIONS/

Sum It Up

Say: Today we estimated how many objects are needed to fill a container.

Ask: How did your estimate compare to the number you counted as you filled a container? [Answers will vary.] /SUMMARIZE/

Extension
Display one container, such as a soup can, and draw it on the board. Have students who used the soup can share how many objects their soup cans held. Record their responses.

Discuss why students' responses vary. [Samples: Different size objects take up different amount of space; same size objects might be placed differently in the space.]

MeasureWorks™ • Grade 1

Name _____

How Many to Fill It?

MeasureWorks

Try This

- Pick one container. Draw it.
- Estimate. How many 🧊 ?
- Write your estimate.
- Count. Write the number.
- Empty the container.
- Repeat with 🧸 .

Quietly count each 🧊 as you drop it in.

Draw	Estimate	Count
❶	_____ 🧊	_____ 🧊
❷	_____ 🧸	_____ 🧸

MeasureWorks™ • Grade 1   Capacity • 1

Compare Capacity Using Less Than or More Than

Planning Your Time
Intro & Demo	Activity	Sum It Up
10 min	15 min	5 min

Objective
Use *more than* and *less than* to compare capacity.

Materials
- Bowls of beans, rice, and unpopped popcorn
- Containers with different capacities
- Foil pans to catch spillage

Grouping
Whole class, then pairs

Open It Up

Note: If you prefer not to use food as a manipulative, use sand or pebbles throughout the activity.

Display two containers with different capacities.

Ask: Will these containers hold the same amount of beans? Will one hold more than the other? Will one hold less than the other? [Answers will vary.]

Hold a container over the foil pan and fill it with beans. Then show how to pour the contents from the filled container to an empty container.

Discuss what students see. Then model using **more than** or **less than** to compare the capacity of the two containers.

Demonstrate & Discuss

Have two volunteers select containers. Have one volunteer fill his or her container with beans.

Have both volunteers display their containers for all to see.

Ask: Will the empty container hold more than or less than the container with beans? [Answers will vary.]

Have the other volunteer pour the beans from the full container into his or her container. *Note:* Be sure the student holds the container over a pan.

Guide students to use the vocabulary *more than* or *less than* to discuss their observations. For example, **say:** Ian's container holds more than Clare's container.

Guide students to see that if a container fills up and beans spill over, then that container holds less than the first container.

Student Activity

Prepare ahead: Each pair will need two containers of different sizes; bowl of beans, rice, or unpopped popcorn; and a foil pan to hold spillage.

Read the directions on the student page aloud to students. Students work in pairs. Each partner selects a container. One partner fills his or her container with beans, rice, or popcorn. Students estimate if the other container will hold more than or less than the filled container. The other partner pours the contents of the filled container into the empty container. Students use *more than* and *less than* to compare the capacities.

Students trade containers with other pairs and repeat to complete the chart.

Informal Assessment

As students work, help them use math language to compare their containers.

Point to one student.
Ask: Does your container hold more than or less than your partner's container? How do you know? [Sample: My container holds more than Ed's container. I know because when I pour my beans into his container, my beans spill into the pan after his container is full.] /SUMMARIZE/

Sum It Up

Say: Today we compared how much two containers can hold.

Ask: Why did you pour beans from one container into another container? [Sample: to compare how much each container holds] /SUMMARIZE/

Does the taller container always hold more? [no] /GENERALIZE/

Literature Connection

The Mysterious Tadpole by Steven Kellogg (Dial, 1977)

Read the story aloud. Have students retell the story. Have them draw pictures to show the tadpole's homes in order by capacity. Then have them retell the story using sentence frames such as the following:
  The sink holds more than / less than the jar.
  The sink holds more than / less than the tub.

Name _____

More or Less?

Try This

Be sure to fill and pour over a pan.

- Pick a 🥫. Compare your 🥫 with your partner's 🥫.
- Estimate, then circle to show which is bigger.
- Fill one 🥫. Pour into the other 🥫.
- Circle to tell what you find.
- Repeat with another 🥫.

MeasureWorks

Circle your estimate.	Circle your findings.
❶ 　　　　　more than My 🥫 will hold　　　　yours. 　　　　　less than	❷ 　　　　　more than My 🥫 will hold　　　　yours. 　　　　　less than
❸ 　　　　　more than My 🥫 will hold　　　　yours. 　　　　　less than	❹ 　　　　　more than My 🥫 will hold　　　　yours. 　　　　　less than

MeasureWorks™ • Grade 1　　　　　　　　　　　　　　　　　　　　　Capacity • 2

Order Containers Using More, Most, Less, and Least

Planning Your Time
- Intro & Demo: 10 min
- Activity: 20 min
- Sum It Up: 5 min

Objective

Compare capacity.
Use *more*, *most*, *less*, and *least* to describe capacity.

Materials

- PopCubes
- Scissors
- Adhesive tape
- Paper squares of different sizes

Grouping

Whole class, then groups of three

Open It Up

Display two containers for students to compare, and then **ask:** Which will hold more PopCubes?

Have students count aloud as you fill each container with PopCubes. Record the number of PopCubes each container can hold.

Point to the container with more as you **say:** This container holds more PopCubes than that container.

Repeat with other containers. Have students use the words *more* or *less* to compare the containers.

Demonstrate & Discuss

Model how to roll a piece of paper to make a paper cone.

Have students count aloud as you fill the cone with PopCubes, then model how to record the capacity on the cone.

Repeat, using paper of a different size to make and fill two more cones. Record the capacity on the cones.

Ask volunteers to hold the cones, then order the cones and model how to use the vocabulary **more, most, less,** and **least** to describe the order. For example, point to the middle cone as you **say**: This cone holds more than the first cone. Then point to the end cone on the right as you **say**: This cone holds the most.

Student Activity

Prepare ahead: Each group will need three paper squares of different sizes, adhesive tape, a pair of scissors, and a collection of PopCubes.

Read the directions on the student page aloud to students. Students work in groups of three. Each student makes a cone-shaped container. Students count as they fill their containers with PopCubes. Students write the number on their cones. Then the members of each group compare the capacities of their cones, using *less*, *more*, and *most* to describe the order. Students draw the three cones and record the number of PopCubes to show the order.

If time allows, have students form new groups and compare the capacity of the cones in the new group. Students use *less* and *least* to describe the order.

Informal Assessment

As students work, be sure they are comparing the number of PopCubes each cone holds and not the physical size of each cone.

Ask: What do the numbers on the cones tell you? [Sample: the number of PopCubes in the cones] /OBSERVE/

How do the numbers on the cones help you know how to order the cones? [Sample: A smaller number means the cone holds less.] /COMPREHENSION/

Sum It Up

Say: Today we compared three containers. We used *more* and *most* or *less* and *least* to describe the comparisons.

Ask: How do you order the cones to show *more*, and then *most*? [Sample: Use the numbers on the cones to go from least to most.] /SUMMARIZE/

Word Study

Fill containers with PopCubes and write the capacity on each container. Write *more, most, less,* and *least* on index cards.

Students select a card, read the word, then select two containers to show *more* and *less*. Then have them choose a third container to show *least* and *most*. Students use the word on the card to describe the containers.

Capacity • T-3   MeasureWorks™ • Grade 1

Name _____

Less, More, the Most

Remember: Compare the number of 🧊 each cone holds—not the sizes of the cones.

Measure Works

Try This
- Roll the paper to make a cone.
- Tape it.
- Fill with 🧊.
- Write how many it will hold.
- Compare with your group.
- Draw to show the order.

Less	More	The Most
Draw the cone.	Draw the cone.	Draw the cone.
_____ 🧊	_____ 🧊	_____ 🧊

MeasureWorks™ • Grade 1

Capacity • 3

Estimate Capacity

Planning Your Time
Intro & Demo	Activity	Sum It Up
10 min	15 min	5 min

Objective
Estimate capacity in nonstandard units. Measure to compare capacity.

Materials
- Beans, rice, or unpopped popcorn
- Collection of plastic containers
- Foil pans to catch spillage

Grouping
Whole class, then groups of four

Open It Up

Display a variety of containers.

Ask: Do you think these containers all hold the same amount of beans? Explain. [Sample: No, some look like they will hold more than the other containers.]

How can we compare the capacity of these containers? [Sample: We can fill one container with beans; then pour the beans from that container to another container to compare.]

Demonstrate & Discuss

Model the student activity. Set out four containers; then estimate to arrange them to show from least to most.

Fill the first container with beans; then pour the beans into the next container. Discuss. Guide students to see that if the second container holds all the beans from the first container, then the order is correct; but if the beans spill over, then the order must be changed.

Continue this process with the other containers. Once the containers are in order, repeat the process to confirm that the order is correct.

Remind students to remove any beans that spilled over from the previous trial and to fill the first container completely before pouring.

Repeat as needed with other sets of containers.

Student Activity

Prepare ahead: Each group will need four containers of different sizes, a bowl of rice, beans, or unpopped popcorn, and a tray to catch spillage.

Read the directions on the student page aloud to students. Students work in groups of four. Each student selects one container. Students estimate and arrange the containers in order from least capacity to greatest capacity.

Students test the order by pouring beans from the smallest container to the next container. If beans spill over, then they change the order of the containers. Students continue in this way to determine the proper order.

Students draw their findings. They repeat the activity with a new set of containers.

Informal Assessment

As students work, be sure that they fill one container completely with beans before they pour the beans into the next container.

Ask: How do you know when to change the order of the containers? [Sample: I must change the order if the beans spill over when I pour, because the container I am pouring into holds less than the one I am pouring from.] /DRAW CONCLUSIONS/

Sum It Up

Say: Today, we measured to show containers in order from least capacity to greatest.

Ask: How do the beans help you measure the capacity of each container? [Sample: The beans show how much one container holds. When I pour the beans into a different container, I am measuring to see how much the next container holds.] /COMPREHENSION/

Extension

Display several containers of similar size.

Ask: If we want to compare the capacities of these containers, is it better to use PopCubes or beans? Why?

Have students fill the containers with PopCubes and compare them. Then have them repeat with beans. Discuss the findings. Elicit that beans are a more precise unit because they are smaller.

Capacity • T-4

MeasureWorks™ • Grade 1

Name _____

How Much?

MeasureWorks

Try This

- Choose a 🫙.
- Compare with your group.
- Estimate. Put the containers in order.
- Fill the first container with beans.
- Pour the beans in the next container.
- Continue. Correct the order.
- Draw to show the correct order.

Clear your work space before each trial.

	Least ──────────────→ **Greatest**			
❶	Draw	Draw	Draw	Draw
❷	Draw	Draw	Draw	Draw

MeasureWorks™ • Grade 1   Capacity • 4

Explore Standard Units of Capacity

Planning Your Time
Intro & Demo	Activity	Sum It Up
10 min	10 min	5 min

Objective
Understand the need for standard units of measure.

Materials
- Bear counters
- Measuring cups
- Clear plastic cups
- Other measuring tools as available (measuring spoons, customary measuring cups, etc.)
- Crayons

Grouping
Whole class, then individuals

Open It Up
Invite three volunteers to each take a handful of bear counters and place them in clear, plastic cups. You should do the same. Set the cups side by side. Compare the number of bear counters in each cup. If the numbers are too close to estimate visually, count the number of bear counters in each cup, record the numbers, and compare.

Discuss why each handful is different. Guide students to see that people have different-size hands, so each handful will be different, depending on the size of the hand.

Demonstrate & Discuss
Display and name the measuring cups and other measuring tools. Then use a 1-cup measuring cup to scoop a cup of bear counters into each of two clear plastic cups. Point out that there are about the same number of bears in each cup.

Ask: How was this measure different from measuring by handfuls? [A cup is always the same amount.]

Point out that each measuring tool measures an exact amount. Explain that a measuring tool that always measures the same amount is called a **standard measure**.

Ask: Do I need to measure an exact amount when:
- I measure flour for a recipe? [yes]
- I measure food for my dog? [no]

Continue with other examples. Have students explain their answers. For each *yes* answer, point out that a standard unit is required.

Student Activity
Prepare ahead: Each student will need crayons.

Read the directions on the student page aloud to students. Students look at and discuss each picture.

Model how to think about the question that goes with each picture by asking: Must I use an exact amount? Then use the words under each picture to finish the question.

Students color the pictures that show when they must use a standard measure.

Informal Assessment
As students are working, be sure they have a clear understanding of what each picture represents.

Point to a picture. **Ask:** Why did you color [or not color] this picture? [Sample: I colored this picture because too much or too little medicine can make you sick, so it is important to take the exact amount.] /DESCRIBE/

Sum It Up
Say: Today we learned about using standard measures.

Discuss each picture and why it should or should not be colored.

Then **ask:** When do you use a standard measure? [Sample: when it is important to know the exact amount] /GENERALIZE/

Book Nook
Knowabout: Capacity by Henry Pluckrose (Franklin Watts, 1988)

Colorful photos provide the foundation for an investigative look at capacity.

Have students answer the questions as you read aloud. Reread the book and focus on math vocabulary. Have students identify words they might use when talking about capacity.

Capacity • T-5   MeasureWorks™ • Grade 1

Name _____

Exact or About?

Try This
- Look at each picture.
- Answer each question.
- Circle yes or no.

Use standard measures when you need an exact amount.

Circle to show when to use standard measures.

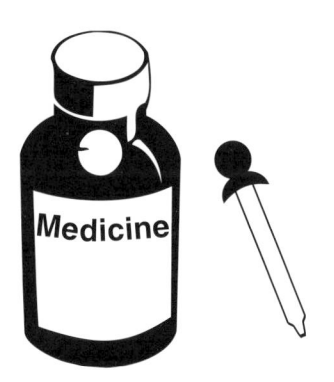

1

Must you use the exact amount when you take medicine?

Yes   No

2

Must you use the exact amount when you have a snack?

Yes   No

3

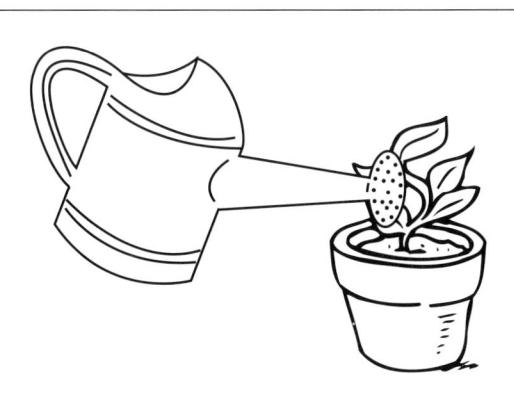

Must you use the exact amount when you water a plant?

Yes   No

4

Must you use the exact amount when you bake muffins?

Yes   No

MeasureWorks™ • Grade 1   Capacity • 5

Explore Math Tools: One Cup

Planning Your Time
Intro & Demo: 10 min
Activity: 20 min
Sum It Up: 5 min

Objective

Explore math tools.
Use a 1-cup measuring cup to measure capacity of containers.

Materials

- Measuring cups
- Clear plastic drinking cups
- Other drinking cups
- Containers (a variety of sizes and shapes, some holding exactly 1 cup)
- Bowls of beans, rice, or unpopped popcorn
- Foil trays to catch spillage

Grouping

Whole class, then groups of four

Open It Up

Display beverage cups in a variety of shapes, sizes, and capacities.

Ask: If a restaurant used these cups to serve milk, would everyone get the same amount? [no]

Fill one large cup with beans, then pour the beans into a second cup with a lesser capacity.

Ask: Do these cups hold the same amount? [no]

Continue in this way to compare the capacity of the cups. Guide students to see that even though all these containers are called cups, they hold different amounts.

Demonstrate & Discuss

Say: There is a standard unit of measure called a **cup.** Its size doesn't change.

Display a 1-cup measuring cup. Tell students that this tool is a measuring cup and that it holds exactly one cup.

Fill the measuring cup with beans, then pour the beans into a clear plastic drinking cup. Repeat, setting the drinking cups next to each other.

Point out how each plastic drinking cup has one cup of beans. Guide students to see that the measuring cup is a standard unit of measure. Point out if a tool is marked *1 cup,* then it will measure one cup.

Discuss when it is important to use a standard unit of measure like a 1-cup measuring cup.

Student Activity

Prepare ahead: Each group will need four containers, a 1-cup measuring cup, a foil tray, and a bowl of beans, rice, or unpopped popcorn.

Read the directions on the student page aloud to students. Students work in groups of four. Each student selects a container. Students draw a picture of each of the four containers.

One student fills a 1-cup measuring cup with beans, sets his or her container in a foil pan, then pours the beans into the container. If the container holds exactly one cup, students color their pictures of that container. If the container holds more than or less than one cup, students mark their pictures with an "X." Each student repeats with his or her container.

Informal Assessment

Ask: Why are you pouring beans from the 1-cup measuring cup into your container? [Sample: to see if the container holds exactly one cup] /COMPREHENSION/

How do you know if a container holds exactly 1 cup? [Sample: All the beans from the 1-cup measuring tool fill my container.] /GENERALIZE/

Sum It Up

Say: Today we explored capacity using the standard measure of one cup. We identified containers that can hold one cup.

Ask: Are all 1-cup containers the same size and shape? [no] /GENERALIZE/

Why not? [Sample: Capacity is how much a container can hold, not its size or shape.] /SYNTHESIZE/

Extension

Discuss the markings on the 1-cup measuring cup. Model good measuring techniques: Place the cup on a flat surface; view contents at eye level; add or remove as needed to measure exactly.

Have students practice filling a 1-cup measuring cup using good measuring techniques.

Capacity • T-6

MeasureWorks™ • Grade 1

Name _____

Color a Cup

Try This

- Fill a [measuring cup] with beans.
- Pick a container.
- Draw it.
- Pour the beans into the container.
- Color it if it holds exactly 1 cup.
- Repeat.
- Draw an "X" on the containers that hold more or less than 1 cup.

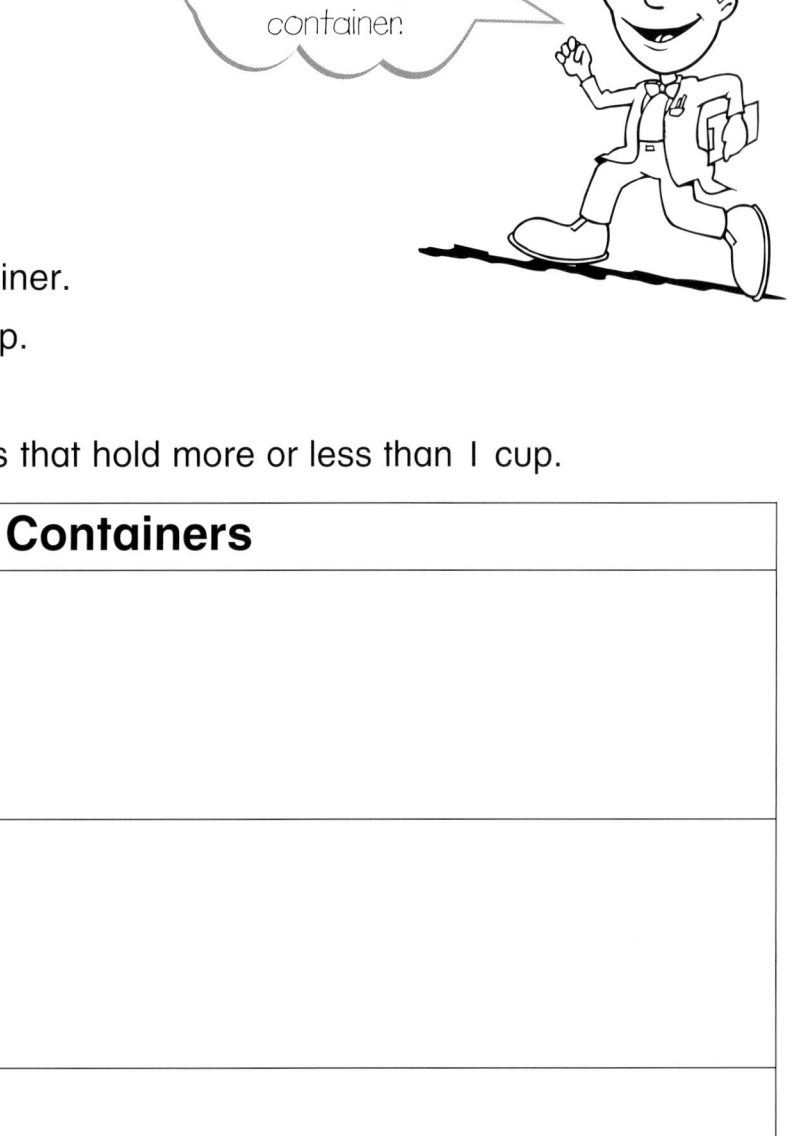

If some beans spill, try to put them back in your container.

Containers
❶
❷
❸
❹

MeasureWorks™ • Grade 1   Capacity • 6

Estimate and Measure One Cup

Planning Your Time
- Intro & Demo: 10 min
- Activity: 15 min
- Sum It Up: 5 min

Objective
Use manipulatives to estimate capacity. Compare estimates to actual measures.

Materials
- Measuring cups
- Containers (various sizes, shapes, capacities—some that hold exactly one cup)
- Bowls of beans, rice, or unpopped popcorn
- Foil trays to catch spillage

Grouping
Whole class, then pairs

Open It Up

Display a 1-cup measuring cup.

Ask: When would you use a 1-cup measuring cup? [Samples: cooking, measuring juice, science projects, etc.]

Explain that measuring cups are used in cooking.

Say: For example, a recipe for cookies might tell you to put in 2 cups of flour.

Ask: If the recipe asked for two containers of flour, what problems might you have? [Sample: I wouldn't know what size container to use. I could end up with cookies that had far too much flour or not enough.] Why is it better to use cups as a unit instead of containers? [Cups are a standard unit; they always hold the same amount.]

Demonstrate & Discuss

Model the student activity. Select a container.

Estimate if the capacity is about one cup.

Fill a 1-cup measuring cup with beans.

Set your container in the foil pan and pour the beans from the 1-cup measuring cup into the container. If beans spill, place them in the container if they will fit.

Discuss the measure. Guide students to see that if the beans fill the container, then the container has a capacity of one cup.

Repeat with other containers so students see the pattern.

Student Activity

Prepare ahead: Each pair will need two containers of different sizes, a bowl of beans, rice, or unpopped popcorn, a 1-cup measuring cup, and a foil tray.

Read the directions on the student page aloud to students. Students work in pairs. Partners select and draw pictures of two containers. Then partners work independently to estimate whether or not the containers hold one cup, and then to measure to check.

If time allows, partners compare results. Did they have the same estimate for each container? Did they have the same result for each container? If students' results for a container do not match, they repeat the measure for that container together.

Informal Assessment

Ask: How do you estimate if your container holds about one cup? [Sample: I think about how much the 1-cup measuring tool holds and I decide if the container will hold the same amount.] /DESCRIBE/

Sum It Up

Say: Today we measured to find which containers hold exactly one cup.

Ask: Did any of the results surprise you? [yes] /OBSERVE/ Why? [Sample: The tall container looks bigger that the 1-cup measuring tool, but it held less than one cup.] /GENERALIZE/

Extension

Tell students that a recipe for sun tea calls for one cup of water and one tea bag. Set one tea bag in $\frac{1}{4}$ cup of water, one in 1 cup of water, and one in 1 quart of water. Have students observe color changes after 1 hour.

Point out that the tea in $\frac{1}{4}$ cup of water is too strong and the tea in 1 quart of water is too weak, but that the tea in 1 cup of water is just right. Discuss why it is important to follow standard measurements in a recipe.

Name _____

About a Cup?

Try This

- Pick two containers.
- Draw them.
- Will your container hold exactly one cup? Circle yes or no.
- Fill a [measuring cup] with beans.
- Pour the beans into your container.
- Circle to show what you find.
- Repeat.
- Compare with your partner.

Be sure to fill the [measuring cup] to the top.

Draw.	Estimate.	Measure.
1	About one cup? Yes No	One cup? Yes No
2	About one cup? Yes No	One cup? Yes No

MeasureWorks™ • Grade 1   Capacity • 7

Explore Math Tools: One Liter

Planning Your Time
Intro & Demo	Activity	Sum It Up
10 min	15 min	5 min

Objective

Explore capacity measurement tools.
Use a 1-liter measuring tool.

Materials

- 1-liter water bottles
- Containers (a variety of sizes and shapes, including extra 1-liter water bottles)
- Pitchers of water
- Large foil pans to catch spillage
- Crayons

Grouping

Whole class, then groups of four

Open It Up

Display a 1-liter water bottle or a 2-liter soda bottle cut in half. Explain that it holds one **liter.**

Fill the 1-liter bottle, then **ask:** How do we describe the capacity of this bottle? [1 liter]

Can other containers hold exactly one liter? [yes]

How can we find out? [Fill the 1-liter bottle with water. Then pour the water into a different container. If the container holds the water exactly, it has a capacity of 1 liter.]

Demonstrate & Discuss

Note: You may wish to roll overhead transparencies into cones to use as funnels.

Fill a 1-liter bottle with water. Then pour the water into a container that holds more than one liter. Repeat with another container. Set the containers next to each other.

Point out that each water bottle contains one liter of water. Guide students to see that the liter is a standard unit of measure.

Discuss when it is important to use a standard unit of measure like a 1-liter measuring tool.

Student Activity

Prepare ahead: Each group will need four containers of different sizes, a pitcher of water, a foil pan, crayons, and a 1-liter water bottle.

Read the directions on the student page aloud to students. Students work in groups of four. Each student selects a container. They draw each of the four containers. Students estimate whether each container holds more than, less than, or the same as 1 liter.

One student fills a 1-liter bottle with water, sets the container in a foil pan, then pours the water into the container. If the container holds one liter, students color their drawings of that container. If the container holds more than or less than one liter, students mark their pictures with an "X." Each student repeats with his or her container.

Informal Assessment

As students work, model how to empty the liter bottle and start the activity over if any water spills into the foil pan.

Ask: Why do you pour water from the 1-liter measuring tool into your container? [Sample: to see if the container holds exactly 1 liter] /COMPREHENSION/
How do you know if a container holds exactly one liter? [Sample: All the water from the 1-liter measuring tool fits in the container.] /GENERALIZE/

Sum It Up

Say: Today we explored capacity, using the standard measure of 1 liter. We found containers that can hold 1 liter.

Ask: Are all 1-liter containers the same size and shape? [no] /GENERALIZE/ Explain. [Capacity is how much something holds, not its size or shape.] /SYNTHESIZE/

Word Study

Explain that an abbreviation is a short way of writing something. Write *1 liter* and *1 L* on the board. Point out that 1 L means 1 liter.

Have students find measurement labels on containers. Write the measurements on the board. Read each measurement aloud, then point to the measurement and have students say it aloud with you.

Name _____

Follow the Liter

Try This

- Fill a 1-liter measuring tool with water.
- Draw a container.
- Estimate: does it hold more than, less than, or the same as 1 liter?
- Pour the water into the container.
- Color the container if it holds exactly 1 liter.
- Write an "X" on the container if it holds more or less than 1 liter.
- Repeat.

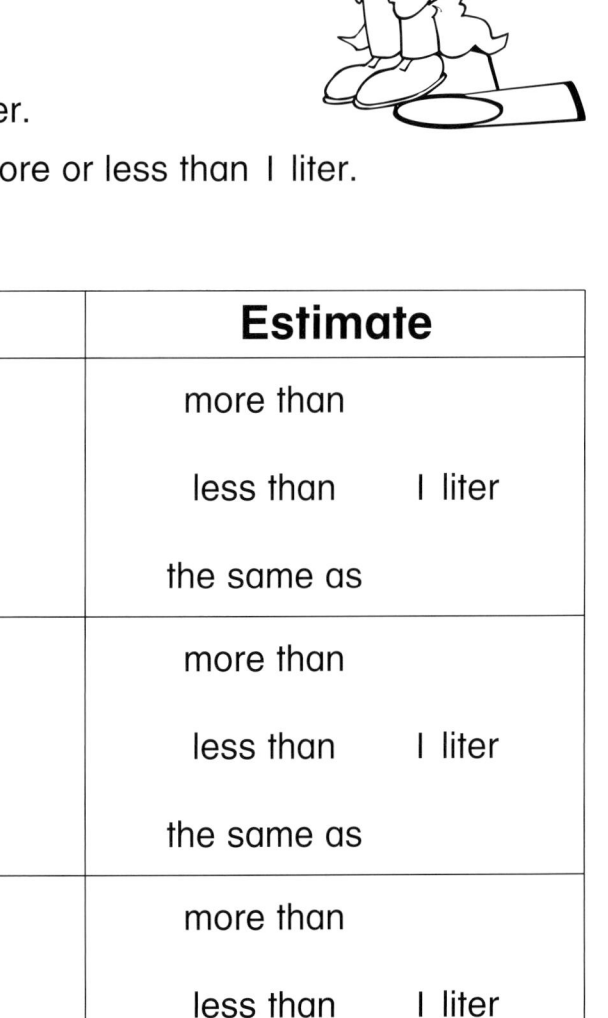

Did you spill? Empty the container. Start over.

Containers	Estimate
①	more than less than 1 liter the same as
②	more than less than 1 liter the same as
③	more than less than 1 liter the same as
④	more than less than 1 liter the same as

MeasureWorks™ • Grade 1

Goldilocks Comes Back

Planning Your Time
Wrap Up: 10 min
Project: 45 min

Objective
Reinforce concepts of capacity.

Materials
- KWL charts
- PopCubes
- Pitchers of water
- Foil trays for spillover
- 3 mugs of different sizes
- 3 bowls of different sizes
- 3 bottles (1 holds exactly 1 liter)
- 1-liter soda bottles
- Crayons

Grouping
Whole class, then individuals

Discuss and Sum It Up
- Have students review their KWL charts.
- As a class, discuss what students knew and what they wanted to learn about capacity.
- Have students share what they learned about capacity.
- Have students update their KWL charts, using what the class shared as well as their ideas.

Assessment
See test on page Assessment • 3.

Project

Note: You may set up this project as a series of measurement stations, or you may provide materials for each group.

Prepare ahead: Collect sets of three mugs in different sizes and sets of three bowls in different sizes. Also collect three bottles. One should be a liter bottle. Label each container in each set with a color so that students can easily distinguish between the items in a set. Make two-sided copies of Capacity • 9–10 and fold into a book for each student. Each student or station will need a 1-liter soda bottle, a collection of PopCubes, crayons, and a pitcher of water.

Read *Goldilocks and the Three Bears* by James Marshall (Dodd Mead, 1992). Then distribute and read "Goldilocks Comes Back" together.

Students perform a series of capacity measurements in order to complete their books. First, students order three mugs from least to greatest capacity. They fill mugs with PopCubes and count the cubes to determine which mug holds the most. They record the number of cubes each holds to show the order of the mugs by capacity.

Next, students use PopCubes to order three bowls from greatest to least capacity. They label to show the order of the bowls by capacity from greatest to least.

Finally, students fill bottles with water and pour it into a 1-liter measuring tool to determine which holds exactly one liter. Students then color the bottle on the page the same color as the bottle that holds exactly one liter.

When students have completed the measurement activities, encourage them to color the pictures in their books.

KWL

I Know	I Want to Know	I Learned
Containers come in different shapes and sizes.	How I can find out how much each container holds?	I can fill each container, then count to see how much each container holds.
I can't tell if two containers hold the same amount just by looking at them.	How can I compare the capacity of the two containers?	I can fill one container, then pour the contents in a second container to see if it holds more, less, or the same.
Math tools, like 1 cup and 1 liter, always measure exactly 1 cup or 1 liter.	How I can use a standard measure to find the capacity of a container?	I can fill a standard measure, then pour the contents in a second container to compare.

Goldilocks Comes Back

This is _____'s book about capacity.

"My bottle holds exactly one liter," said Goldilocks. "Which one is it?"

Pour and measure. Color the bottle to match.

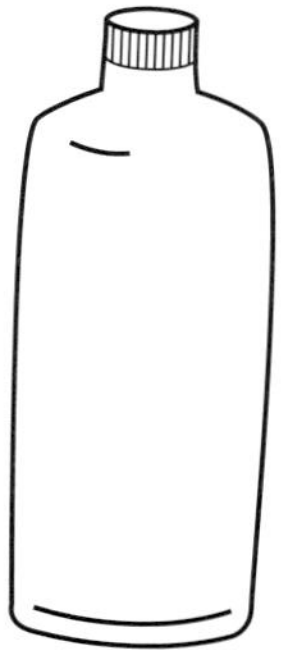

My Book page 3

MeasureWorks™ • Grade 1 Capacity • 9

"My mug holds the most," said Papa Bear.
"My mug holds more," said Mama Bear.
"My mug holds less," said Baby Bear.
"Someone MIXED THEM UP."

Fill each mug with cubes.
Write how many cubes fit in each mug.

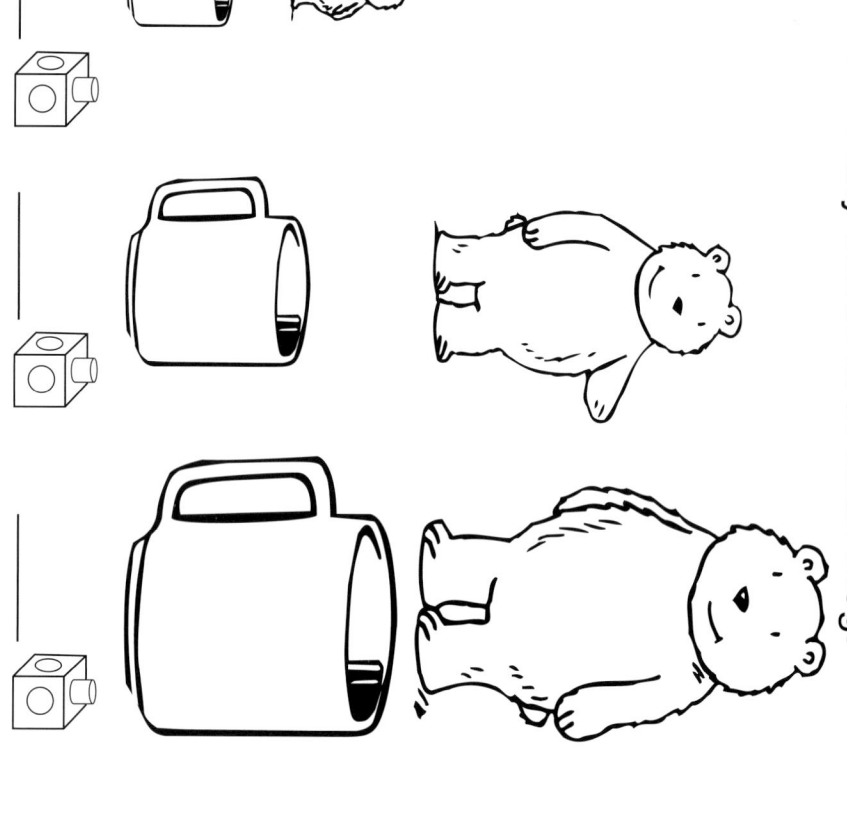

My Book page 1

---

fold

"My bowl holds the most," said Papa Bear.
"My bowl holds less," said Mama Bear.
"My bowl holds the least," said Baby Bear.
"Someone MIXED THEM UP."

Fill each bowl.
Write how many cubes fit in each.

My Book page 2

Capacity • 10

MeasureWorks™ • Grade 1

Notes

Answer Key for Capacity Unit

Capacity • 5
1. yes
2. no
3. no
4. yes

Measure Works
Weight

Unit Introduction

In this unit, students

☞ Explore concepts of weight.
☞ Discover that size and weight are different measures.
☞ Measure the weight of common objects.
☞ Estimate weight related to 1 pound and 1 kilogram.

Assessment
A unit test in multiple-choice format is provided on page Assessment • 4.

KWL

Use a KWL chart to activate prior knowledge and set learning goals as a class. A reproducible KWL chart is provided on page BLM • 7.

Have students keep the KWL chart in their math folders and add to it as they work through this unit.

Games for Practice and Review
Use the MeasureWorks Game Board to reinforce learning. Game rules begin on page BLM • 13.

Focus on Vocabulary

about the same (p. T-5)	heaviest (p. T-2)	lightest (p. T-2)	the same (p. T-1)
close to (p. T-5)	kilogram (p. T-6)	balance (p. T-1)	unit of weight (p. T-3)
heavier (p. T-1)	lighter (p. T-1)	pound (p. T-4)	weigh (p. T-3)

Write each vocabulary word on an index card. Make a display to refer to as students work through the unit. Encourage students to build vocabulary awareness by awarding points each time students use one of the vocabulary words.

Heads Up!
Students frequently assume that size and weight are the same and that if one object is bigger than another, it is also heavier. The two-pan balance allows students to compare weight while visually comparing size. They discover that sometimes smaller objects are heavier than bigger objects.

Be sure students place objects being weighed in the middle of each pan. Placing objects in different positions affects the accuracy of the measure.

Book Nook
Who Sank the Boat?

by Pamela Allen

1982: Coward, McCann, Inc.

A delightful poem about balance and weight. See Connection on page Weight • T-2.

Weight

Manipulatives

Pages	Learning Goals	Balances	Bear counters	Kilogram masses	Standard mass sets
T-1–1	Explore math tools. Use a balance to weigh common objects.	✔			
T-2–2	Estimate *heavier* and *lighter*. Use a balance to check predictions.	✔			
T-3–3	Estimate and weigh objects using nonstandard units.	✔	✔		
T-4–4	Predict and test to identify objects that weigh more and less than one pound.	✔			✔
T-5–5	Identify objects that weigh about 1 pound. Decide whether an object weighs close to 1 pound.	✔	✔		✔
T-6–6	Predict and test to identify objects that weigh more or less than 1 kilogram.	✔		✔	

MeasureWorks™ • Grade 1

Lighter, Heavier, the Same

Planning Your Time
Intro & Demo	Activity	Sum It Up
10 min	15 min	5 min

Objective
Explore math tools.
Use a balance to weigh common objects.

Materials
- Balances

Grouping
Whole class, then four small groups

Open It Up

Give each student an object. Have students turn to a partner and predict which of them has the heavier item and which has the lighter item.

Display a balance. Tell students that this tool is called a **balance.** Have students observe the balance at rest and talk about what they see.

Demonstrate & Discuss

Have two volunteers show the class their objects and share their prediction about which is heavier and which is lighter.

Then have each student place the object in a pan of the balance. Observe the results. Introduce vocabulary: **lighter, heavier,** and **the same.**

Ask: Is one pan higher? Why? [Yes, because that object is lighter.] How do you know which object is heavier? [The side with the heavy object goes down.] What does it mean when the sides are even? [The objects weigh the same.] How can we make the sides of the balance even? [Put more little things in the light side to make it heavier.]

Repeat with other pairs of objects.

Student Activity

Prepare ahead: Each group will need a balance.

Read the directions on the student page aloud to students. Students work in groups, with one balance per group. They choose two objects, place one in each pan on the balance and decide which object is heavier and which is lighter. They draw a picture of each object to record their work.

Help students complete each sentence by drawing the objects they measured in the spaces provided.

Informal Assessment

As students work, encourage them to use the vocabulary: *lighter, heavier,* and *the same.*

Ask: What do you notice about the pans? [Sample: The side with the heavier object always goes down.] /OBSERVE/

Why do you think one side is higher (or lower)? [because that object is lighter (or heavier)] /CAUSE AND EFFECT/

Do you think bigger objects always make the balance go lower? [no] Why or why not? [Big objects can sometimes be light.] /INFER/

Sum It Up

Say: Today we learned how a balance helps us compare weights. We learned to tell which object is heavier and which object is lighter.

Ask: How does a balance work? [Sample: The heavy side goes down; the light side goes up.] /DESCRIBE/

Is the bigger object always heavier? [No. You need to compare their weights to decide.] /INFER/

Summarize findings in a class list or display.

Bulletin Board Idea
THINGS WE WEIGHED
These are heavier than 5 PopCubes.

Name _____

Lighter or Heavier

MeasureWorks

Try This

- Pick two objects.
- Put one object in each pan.
- Show what happens to the pans.
- Draw a picture. Write sentences.
- Repeat with other objects.

Draw one object on each side of the balance.

Draw.	Write.
1	_____ is heavier than _____ . _____ is lighter than _____ .
2	_____ is heavier than _____ . _____ is lighter than _____ .
3	_____ is heavier than _____ . _____ is lighter than _____ .

MeasureWorks™ • Grade 1

Weight • 1

Light, Lighter, Lightest, Heavy, Heavier, Heaviest

Planning Your Time
Intro & Demo	Activity	Sum It Up
10 min	15 min	5 min

Objective

Estimate *heavier* and *lighter*.
Use a balance to check predictions.

Materials

- Balances

Grouping

Whole class, then four small groups

Open It Up

Hold up two objects for students to compare.

Ask: Which object is bigger? How can you tell? [Answers will vary.]

Which object do you think is heavier? Why? [Answers will vary.]

Invite a volunteer to hold one item in each hand and tell which feels heavier (or lighter).

Demonstrate how to check predictions using a balance.

Demonstrate & Discuss

Have two students each pick an object. Have them compare the weight by holding the objects and predict which is heavier and which is lighter. Let them check their ideas using a balance. Record findings.

Now add a third object. Guide students to use *lighter than* and *heavier than* to compare the three objects.

Order the objects from **heaviest** to **lightest.**

Guide students to use the vocabulary. Set sentence frames such as:

"The [object] is lighter than [second object]."

"The [object] is heavier than [last object]."

Student Activity

Prepare ahead: Each group will need a balance.

Read the directions on the student page aloud to students. Students work in groups of three. They compare three objects by weight and order the three objects from light to lighter to lightest.

Help students understand that heavier and lighter are relative terms. An object that is light can be heavier than another light object.

Informal Assessment

As students work, encourage them to use the vocabulary: *lighter, heavier, lightest,* and *heaviest.*

Ask: How do you use the balance to find out which object is heavier than another? [Sample: I put one object in each pan. The object in the pan that goes down is heavier.] /DESCRIBE/

Sum It Up

Say: Today we compared three light objects to see which was light, which was lighter, and which was lightest.

Ask: How did your estimates compare to what you saw when you used the balance? [Sample: My estimates improved as I practiced.] /SUMMARIZE/

Book Nook

Who Sank the Boat? by Pamela Allen (Coward, McCann, Inc., 1982)

Have students draw a picture of their favorite story character. Then have them present their pictures. Compare characters by weight. Discuss which might be heavier, heaviest, lighter, lightest or the same. Have students use the illustrations to support their conclusions. Discuss why.

Name _____

Light, Lighter, Lightest, Heavy, Heavier, Heaviest

MeasureWorks

Use a balance to check your order.

Try This
- Pick three objects.
- Compare their weight.
- Put them in order.
- Draw what you found out.

Draw.

❶ light	❷ lighter	❸ lightest
heaviest	heavier	heavy

Complete each sentence.

❹ _____ is lighter than _____ .

❺ _____ is lighter than _____ .

❻ _____ is heavier than _____ .

❼ _____ is heavier than _____ .

MeasureWorks™ • Grade 1   Weight • 2

Weigh with Nonstandard Units

Planning Your Time
Intro & Demo	Activity	Sum It Up
10 min	10 min	5 min

Objective
Estimate and weigh objects using nonstandard units.

Materials
- Balances
- Bear counters
- Tennis ball

Grouping
Whole class, then four small groups

Open It Up

Display a balance. Put an object such as an eraser in one pan and a bear counter in the other.

Ask: How much does this object **weigh**? [Sample: You can't tell because the pans don't balance.]

Explain that to weigh things, we need to agree on an unit of weight.

Say: Today, we are going to weigh things. Our **unit of weight** is the bear counter.

Demonstrate & Discuss

Display the balance, a tennis ball, and some bear counters.

Invite students to predict the number of counters needed to balance the tennis ball. Record.

Have volunteers check their predictions. Mark by each prediction "too many," "too few," or "just right."

Repeat with another object, such as a marker.

Model how to record estimates and actual measures.

Student Activity

Prepare ahead: Each group will need a balance and a collection of bear counters.

Read the directions on the student page aloud to students. Students work in groups. They take turns selecting an item and estimating its weight in bear counters. They then weigh the object, and compare its actual weight to the estimate.

Informal Assessment

As students work, help them use math language to describe what they are doing and thinking.

Ask: What unit of weight are you using? [bear counters] /COMPREHENSION/

What do you notice about two of the objects you measured? [Sample: They don't weigh the same.] /OBSERVE/

Which is heavier/lighter? [The object that needs more bear counters is heavier. The one that needs fewer bear counters is lighter.] /GENERALIZE/

What do the numbers on your recording sheet tell you? [how many counters each object weighs] /COMPREHENSION/

Sum It Up

Say: Today we estimated and weighed things using bear counters as our unit of weight.

Ask: Which was your best estimate? How do you know? [Examples will vary. Elicit that the best estimate is the one that is closest to the actual number of bears that balanced the scale.] /INFER/

Which of the objects you measured was the *heaviest*/the *lightest*? How do you know? [Sample: The stapler was the heaviest because it weighed the most bear counters. The marker was the lightest because I needed only a few bear counters to balance it.] /COMPARE AND CONTRAST/

Science Connection

The weight of an object is the force of gravity pulling the object down toward Earth. Weight measures the effect of gravity on the object's mass (the quantity of material it contains). There is no gravity in space to pull people down. In space, away from the Earth's gravity, astronauts are so light that they float. [Source: *The Story of Weights and Measures* by Anita Ganeri (Oxford University Press, 1996) p. 14.]

Name _____

Weigh with Bears

Measure Works

Try This

- Find an object.
  Draw a picture of it.
- Estimate. About how many 🐻 will balance your object?
- Write your estimate.
- Measure.
- Write what you find out.

Count the 🐻 quietly as you put them in the pan.

What I Measured	Estimate	Measure
❶	_____ 🐻	_____ 🐻
❷	_____ 🐻	_____ 🐻
❸	_____ 🐻	_____ 🐻
❹	_____ 🐻	_____ 🐻

Which object was the heaviest? _____

How do you know? _____

MeasureWorks™ • Grade 1   Weight • 3

More or Less than 1 Pound

Planning Your Time
Intro & Demo: 10 min
Activity: 20 min
Sum It Up: 5 min

Objective
Predict and test to identify objects that weigh more and less than one pound.

Materials
- Balances
- Standard mass sets

Grouping
Whole class, then four small groups

Open It Up

Teaching Tip: Have students sit in a circle. Place a collection of classroom objects in the middle of the circle. As students take their turn holding the 1-pound weight, have them visually pick an object that might weigh more or less than one pound.

Say: Today we are going to compare the weight of classroom objects to one **pound**. We will make a list of things that weigh more than one pound and less than one pound.

Pass around a 1-pound weight. Tell students that this object weighs about one pound. Explain that a pound is a unit of weight. Discuss which classroom objects might weigh about the same as the 1-pound weight.

Demonstrate & Discuss

Teaching Tip: You may find it easier to measure items if you remove the pans on the balance and place the items directly on the platforms. Make sure you remove both pans to keep the balance accurate.

Pass around a classroom object.

Ask: Do you think this weighs more than 1 pound or less than 1 pound? [Answers will vary.] Record students' predictions.

Ask: How can we find out? [Elicit that you can put the 1-pound weight in one pan and the object in the other. Then observe which is heavier.]

Demonstrate this process. Discuss the position of the pans and what the balance shows. Model how to record findings on the student page.

Student Activity

Prepare ahead: Each group will need a balance and a 1-pound weight.

Read the directions on the student page aloud to students. Students work in groups. Students predict whether an object weighs more or less than one pound. They use a balance to check.

Students use pictures or words to record their findings.

Informal Assessment

As students work, help them use math language to describe what they are doing and thinking.

Ask: What unit of weight are you using? [pounds] /COMPREHENSION/

What object do you think weighs more (or less) than 1 pound? [Answers will vary.] /PREDICT/

How can you find out if you are right? [Sample: I can put a 1-pound weight in one pan and my object in the other. If the object weighs more than 1 pound, it will make the pan go down. If it weighs less than 1 pound, it will go up.] /DESCRIBE/

Sum It Up

Say: Today we identified everyday objects that weigh more than 1 pound and less than 1 pound.

Ask: What things did you find that weigh more than (or less than) 1 pound? [Answers will vary.] /SUMMARIZE/

Word Study

Write the words *heavier* and *lighter* on the board.

Say: An object that weighs more than 1 pound is heavier than 1 pound. An object that weighs less than 1 pound is lighter than 1 pound.

Encourage students to complete sentence frames such as these:
The _____ weighs _____ (more/less) than 1 pound.
The _____ is _____ (heavier/lighter) than 1 pound.

Name _____

More or Less than 1 Pound

MeasureWorks

Try This

- Choose an object.
- Predict. Will it weigh more or less than ☐ ? 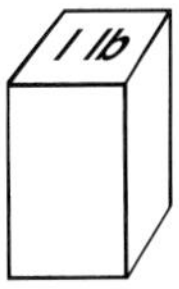
- Check.

  Use the ⚖ .

- Record what you find out.
- Repeat.

Put a 1-pound weight in one pan. Put an object in the other. Compare.

More Than 1 Pound	Less Than 1 Pound
❶	❷
❸	❹

MeasureWorks™ • Grade 1   Weight • 4

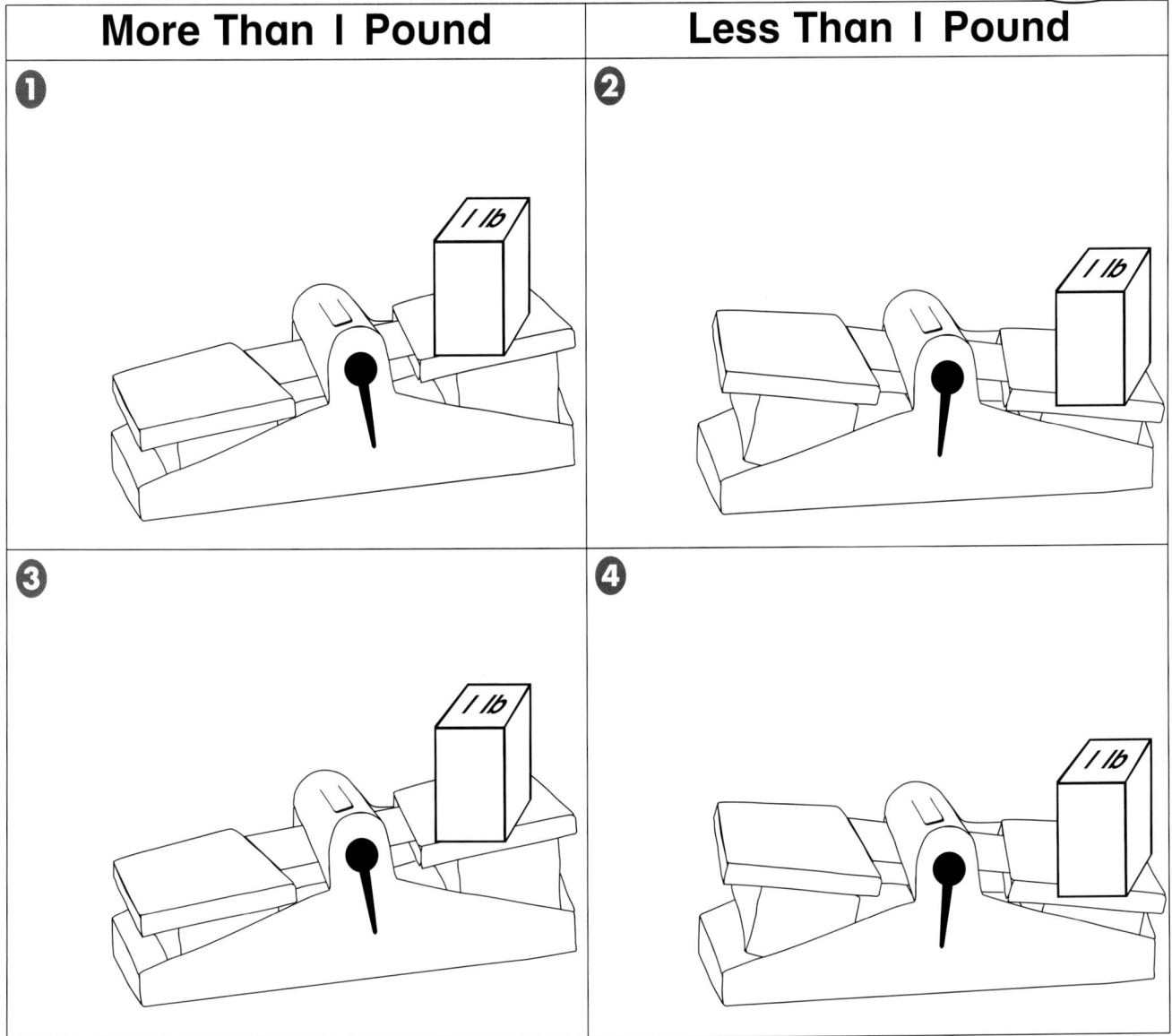

About 1 Pound

Planning Your Time
- Intro & Demo: 10 min
- Activity: 15 min
- Sum It Up: 5 min

Objective

Identify objects that weigh about 1 pound. Decide whether an object weighs close to 1 pound.

Materials

- Balances
- Standard mass sets
- Bear counters
- 1-pound packages of foods such as pasta, rice, crackers, etc.

Grouping

Whole class, then four small groups

Open It Up

Have students sit in a circle.

Have students pass around a package that contains about 14 ounces of cereal, crackers, pasta, or other food.

Ask: Does this package feel lighter than 1 pound, heavier than 1 pound, or **about the same?** [Answers will vary.] How can we find out? [Weigh the package.]

Demonstrate & Discuss

Put a 1-pound weight in one pan of the balance and the package in the other.

Ask: How can you tell if the package weighs exactly 1 pound? [The balance should be even.] How could we use bear counters to make the pans balance? [Add counters one at a time to the lighter object until the pans balance.]

Add bear counters to the lighter side of the balance. (*Note:* It will take about 8 bears (2 oz) to balance the scale.)

Say: I used fewer than 10 bear counters to balance the pans. That means that our package weighs **close to** 1 pound.

Repeat with another package that weighs about 8 ounces. After you add 15 bear counters, stop.

Say: It will take a lot of bear counters to balance the pans. That means that the package does not weigh close to, or about, 1 pound.

Repeat with other objects.

Student Activity

Prepare ahead: Each group will need a balance, 10 bear counters, and a 1-pound weight.

Read the directions aloud. Ask students to summarize how the activity works. Students work in small groups to predict which everyday objects weigh about 1 pound. They draw the objects. Then they place each object on a balance with a 1-pound weight on the other side. They circle whether the object weighs more than, less than, or the same as 1 pound. For objects that weigh more or less than 1 pound, students use bear counters to try to balance the object and record how many bears they used. If the pans balance with fewer than 10 bear counters, they record that the object weighs about 1 pound.

Informal Assessment

As students work, help them use math language to describe what they are doing and thinking.

Ask: How many bear counters did you need to balance the pans? [Answers will vary.] /OBSERVE/

What does it mean if the pans still aren't level after you use all your bear counters? [It means that the object weighs a lot more or a lot less than 1 pound.] /INFER/

Sum It Up

Say: Today we looked for objects that weighed about 1 pound.

Ask: When you use the balance, how do you know whether your object weighs about 1 pound? [An object weighs about 1 pound if I can balance it with the 1-pound weight or if I can balance it with the 1-pound weight and a few bear counters.] /SUMMARIZE/

Extension

Pick small items such as crayons or clothespins.

Ask: How many crayons do you think there are in one pound?

Record student's estimates.

Then have students test their estimates with a balance and share what they find out.

Weight • T-5

MeasureWorks™ • Grade 1

Name _____

About 1 Pound

Try This

- Draw something in your classroom you think weighs about 1 pound.
- Put it on the ____.
- Circle more than, less than, or the same as 1 pound.
- Add 🐻. Try to make the pans balance.
- Record how many 🐻 you used.
- Did your object weigh close to 1 pound? Circle Yes or No.

If the pans are level, your object weighs 1 pound.

Draw.	Circle.	How many bears balance the scale?	Close to 1 pound?
❶	more than less than the same as 1 pound	_____	Yes No
❷	more than less than the same as 1 pound	_____	Yes No
❸	more than less than the same as 1 pound	_____	Yes No

MeasureWorks™ • Grade 1   Weight • 5

More or Less than 1 Kilogram

Planning Your Time
Intro & Demo: 10 min
Activity: 15 min
Sum It Up: 5 min

Objective
Predict and test to identify objects that weigh more or less than 1 kilogram.

Materials
- Balances
- Kilogram masses

Grouping
Whole class, then four small groups

Open It Up

Teaching Tip: Have students sit in a circle. As they wait for their turn, help them listen as each classmate compares the objects by weight.

Pass around a kilogram mass and a package that contains about one kilogram of rice (about 2 pounds). Explain that the mass is a unit of measure called a **kilogram.**

Encourage each student in turn to close his or her eyes and act out being a balance.

Ask: Does this package feel lighter than 1 kilogram, heavier than 1 kilogram, or about the same as 1 kilogram? How did you decide? [Answers will vary.] How can we find out? [Elicit that we can weigh the package.]

Demonstrate & Discuss

Invite a volunteer to select an object in the classroom. Pass the object around. Ask students to estimate whether it weighs more or less than 1 kilogram.

Ask: How can you check? [We can put the 1-kilogram mass in one pan and see if the object balances.]

Model how to check using the balance. Discuss the position of the pans and what the balance shows. Model how to record results on the student page.

Student Activity

Prepare ahead: Each group will need a balance and a kilogram mass.

Read the directions on the student page aloud to students. Students work in groups. They predict which everyday objects might weigh more than or less than one kilogram. Then they verify their predictions.

Help students use words and pictures to record their findings.

Informal Assessment

As students work, help them use math language to describe what they are doing and thinking.

Ask: What unit of weight are you using? [kilogram] /COMPREHENSION/

What object do you think weighs more (or less) than 1 kilogram? [Answers will vary.] /PREDICT/

How can you find out if you are right? [Sample: I can put a 1-kilogram weight in one pan and my object in the other. If the object weighs more than 1 kilogram, it makes the pan go down. If it weighs less than 1 kilogram, it goes up.] /DESCRIBE/

Sum It Up

Say: Today we found objects that weigh more than or less than 1 kilogram.

Have students share their findings.

Ask: What things did you find that weigh more than (or less than) 1 kilogram? [Answers will vary.] /SUMMARIZE/

Science Connection

Discuss the ways rocks are alike and different (size, color, shape, texture, weight, etc.).

Then arrange for a "kilogram rock hunt." Let students search for rocks in the schoolyard or nearby neighborhood.

Teaching Tip: If this is not possible, provide a collection of rocks for students to use in the classroom. Have each student select one rock to share with the class and describe various qualities of the rock, including whether it is heavier than, lighter than, or about 1 kilogram.

Name _____

Kilogram Crush

Try This

- Choose an object.
- Predict. Does it weigh more or less than ⬜?
- Check. Use a balance.
- Record what you find out.
- Repeat.

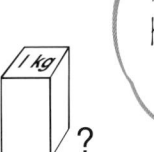

A kilogram is heavier than a pound.

Draw the object.	Predict.	Check. Draw the ⬜ and the object.
1	more than less than ⬜	
2	more than less than ⬜	

MeasureWorks™ • Grade 1   Weight • 6

Read and Write About Weight

Planning Your Time

Wrap Up	Project
15 min	30 min

Objective
Summarize learning about weight.

Materials
- KWL charts
- Crayons
- Scissors

Grouping
Whole class, then individuals

Discuss and Sum It Up
- Have students review their KWL charts.
- As a class, discuss what they knew about weight and what they wanted to learn.
- Have volunteers tell what they learned. Make a list of their ideas.
- Have students complete the third column of their charts. Encourage them to write their own ideas or copy ideas from the class list.

Assessment
See test on page Assessment • 4.

Project

Prepare ahead: Each student will need a pair of scissors and crayons.

Students make and write a take-home book about weight. The book reinforces the idea that big things can be lighter than small things.

Small title page: Read the title of the story book. Discuss what the first picture shows. Elicit that the picture shows a boy and a girl, a beach ball, a shell, and a sail boat. Have students compare the objects by size. Then help them complete each sentence to describe the objects.

Small page 1: Discuss what the picture shows. Elicit that the ball and shell are on a balance. The balance shows that the ball is lighter than the shell. Help students complete each sentence to compare the objects shown.

Continue in the same way for small pages 2–5.

If time permits, invite students to color the pictures in their books. Have students take the books home to share with their families.

KWL — Know • Want to Know • Learned

I Know	I Want to Know	I Learned
Things can be heavy or light.	How many books can I carry?	I can measure weight in kilograms or pounds.
Light things are easy to carry.	How much do 10 counters weigh?	Ten counters weigh the same as a stapler.
I get weighed when I go to the doctor.	How heavy is a pound?	Lots of things weigh about one pound.

Write Yes or No.

Is the boy heavier than the boat? _____

Is the boy lighter than the boat? _____

Do they weight about the same? _____

Finish each sentence.
Write heavier or lighter or the same weight.

The boy is _____.

The boat is _____.

My Book page 5

Light and Heavy

This is _____'s book about weight.

MeasureWorks™ • Grade 1　　　　　　Weight • 7

Write Yes or No.

Is the boy heavier than the girl? _____

Is the boy lighter than the girl? _____

Do they weight about the same? _____

Finish each sentence.

Write heavier or lighter or the same weight.

The boy is _____ as the girl.

The girl is _____ as the boy.

My Book page 4

Write Yes or No.

Is the ball bigger than the shell? _____

Is the ball heavier than the shell? _____

Finish each sentence. Write heavier or lighter.

The ball is _____.

The shell is _____.

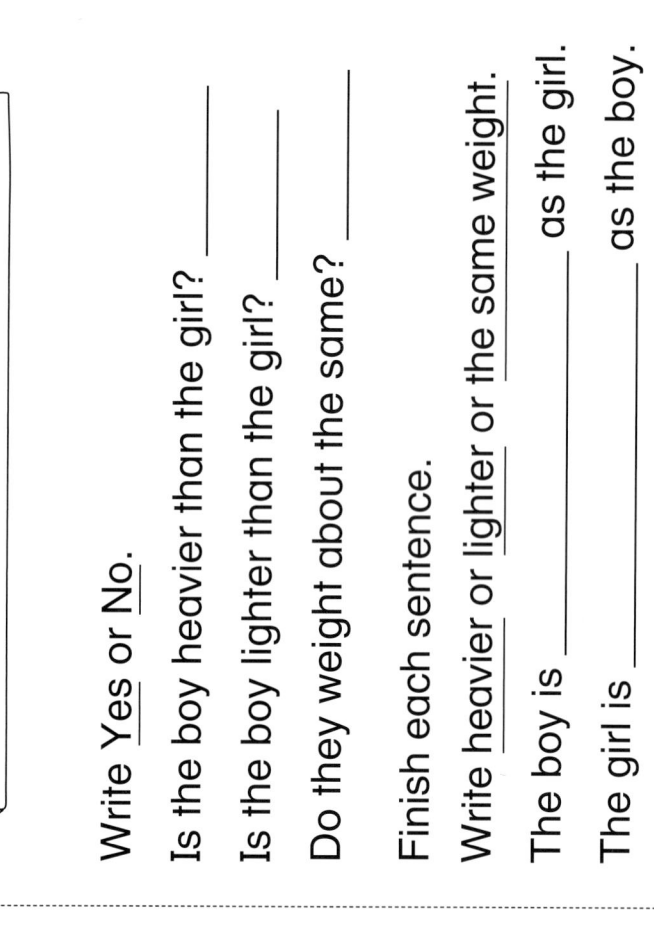

Finish each sentence. Write ball or shell.

The _____ is lighter.

The _____ is heavier.

My Book page 1

Write Yes or No.

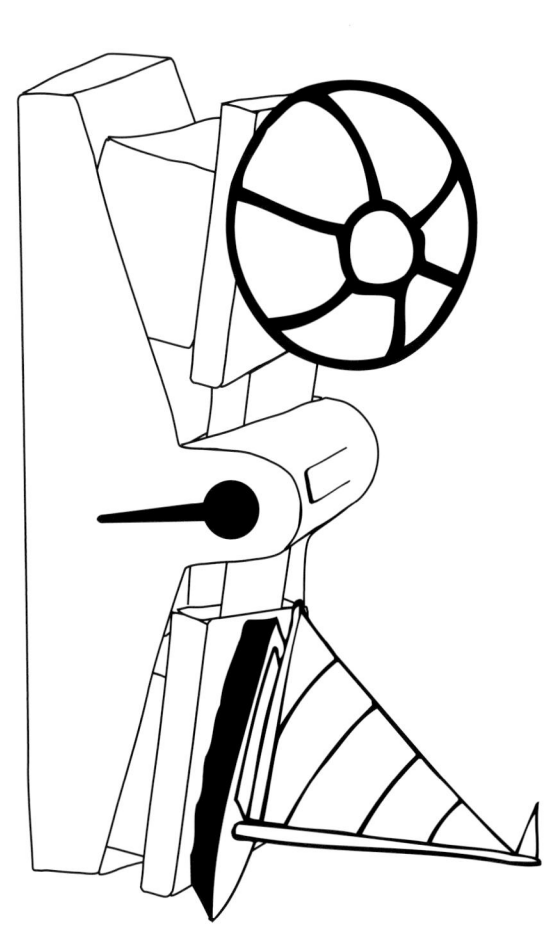

Is the ball bigger than the boat? _____

Is the ball heavier than the boat? _____

Finish each sentence. Write heavier or lighter.

The ball is _____.

The boat is _____.

Finish each sentence. Write ball or boat.

The _____ is lighter.

The _____ is heavier.

My Book page 3

Write Yes or No.

Is the shell bigger than the boat? _____

Is the shell heavier than the boat? _____

Finish each sentence. Write heavier or lighter.

The shell is _____.

The boat is _____.

Finish each sentence. Write shell or boat.

The _____ is lighter.

The _____ is heavier.

My Book page 2

MeasureWorks™ • Grade 1 Weight • 9

Answer Key for Weight Unit

Weight • 7
   5. yes
      no
      no
      heavier
      lighter

Weight • 8
   1. yes
      no
      lighter
      heavier
      ball
      shell
   4. no
      no
      yes
      the same weight
      the same weight

Weight • 9
   3. yes
      no
      lighter
      heavier
      ball
      boat
   4. no
      no
      lighter
      heavier
      shell
      boat

MeasureWorks™
Temperature

Unit Introduction

In this unit, students

☞ Explore concepts of temperature.
☞ Compare and order items of different temperatures.
☞ Read a thermometer with a Fahrenheit scale.
☞ Estimate temperatures.

Assessment
A unit test in multiple-choice format is provided on page Assessment • 5.

KWL

Use a KWL chart to activate prior knowledge and set learning goals as a class. A reproducible KWL chart is provided on page BLM • 7.

Have students keep the KWL chart in their math folders and add to it as they work through this unit.

Games for Practice and Review
Use the MeasureWorks Game Board to reinforce learning. Game rules begin on page BLM • 13.

Focus on Vocabulary

cold (p. T-2)	cooler (p. T-4)	Fahrenheit (p. T-5)	hottest (p. T-4)	warm (p. T-2)
colder (p. T-4)	coolest (p. T-4)	high (p. T-3)	low (p. T-3)	warmer (p. T-4)
coldest (p. T-4)	degrees (p. T-5)	hot (p. T-2)	temperature (p. T-1)	warmest (p. T-4)
cool (p. T-2)	estimate (p. T-6)	hotter (p. T-4)	thermometer (p. T-2)	

As you introduce the vocabulary words, write them on index cards. You can hang the classroom thermometer in the center of a bulletin board. Then display the temperature vocabulary words around the thermometer.

Heads Up!
Students may not be as familiar with measuring temperature as they are with measuring length or weight. They may have little idea of how warm or cold a temperature such as 70°F is.
Allow plenty of time to discuss the meaning of temperature, why temperature is important, and how temperature is measured. Make sure students understand that weather words, such as *sunny* and *rainy*, do not necessarily describe temperature.

Book Nook
On the Same Day in March: A Tour of the World's Weather
by Marilyn Singer
2000: Harpercollins

This book describes 17 locations around the world on the same day in March. As you read, help students compare temperatures, from the cold Arctic region to sunny Barbados.

Temperature

Manipulatives

Pages	Learning Goals	Classroom thermometer	Sliding thermometers
T-1–1	Explore temperature. Discuss temperature in everyday situations.		
T-2–2	Review words related to temperature.	✓	
T-3–3	Compare temperatures. Order items according to temperature.		
T-4–4	Compare and order temperatures. Use temperature vocabulary.		
T-5–5	Discover how a thermometer works. Read a thermometer (Fahrenheit scale).	✓	✓
T-6–6	Estimate water temperatures. Model and write temperatures.	✓	✓
T-7–7	Practice reading a thermometer. Connect seasons and activities to temperature.	✓	✓

Get to Know Temperature

Planning Your Time

Intro & Demo	Activity	Sum It Up
10 min	10 min	5 min

Objective
Explore temperature.
Discuss temperature in everyday situations.

Materials
- none -

Grouping
Whole class, then individuals

Open It Up

Ask: What is **temperature?** [Temperature is how hot or cold something is.]

Play "Give Me a Clue." Tell the class that you are going to describe something in your home. Ask students to carefully listen to hear all the clues.

Say: It is about this tall (show about 10 inches with your hands). It has a colored label on the outside. It weighs about as much as four apples. A white liquid is inside. I keep it cold in the refrigerator. [carton of milk]

Ask: What was the only clue about temperature? [keeping it cold in the refrigerator]

Demonstrate & Discuss

Say: Temperature is important in many situations. Share about some times when you noticed temperature. [Samples: I noticed that it was cold this morning; an oven is hot; my dad took my temperature when I was sick; the refrigerator keeps food cold so it doesn't spoil.]

Make a list of students' responses on the board.

Ask: When is it important to know the temperature of something? [Samples: It's important to know the temperature when you cook so that the food turns out right and does not burn; a high body temperature (fever) means you are sick; knowing the outside temperature helps us plan what to do and what to wear.]

Student Activity

Read the directions on the student page aloud to students. Students read the two sentences in each exercise, and circle the sentence that relates to temperature. If you anticipate that the vocabulary will be too difficult for students, read the sentences aloud to the class as students circle the answers.

Informal Assessment

As students work, help them think about concepts and words related to temperature.

Ask: Is a sunny day always warm? [no] /OBSERVE, RECALL/

Why isn't the sentence about having a cold about temperature? [Sample: because the word *cold* in this case is the name of a sickness] /COMPREHENSION/

Besides "boiling," what words tell us about the temperature of water? [Sample: hot] /RECALL/

Sum It Up

Say: Today we talked about what temperature is and how it is important.

Ask: How was temperature important to you today? [Samples: It was cold, so I had to wear mittens and boots. My dad made eggs for breakfast, but they got cold and tasted bad. I was really hot when we came in from playing outside.] /ANALYZE/

Science Connection

Temperature is important to birds, too. Most birds sit on their eggs to keep them warm. Sometimes a bright light bulb is used to keep chicken eggs warm until they hatch. People often think birds migrate to keep warm, but actually, most birds migrate to find foods that become scarce in cold weather.

Name _____

Temperature or Not?

Temperature is how hot or how cold something is.

Try This

- Read the two sentences.
- One tells about temperature.
- Circle the sentence that tells about temperature.

Circle the sentence about temperature.

1

A. This room is big.

B. This room is warm.

2

A. The sprinkler is making me wet.

B. The sprinkler is making me cool.

3

A. Set the timer for 45 minutes.

B. Turn on the oven to warm it up.

4

A. She has a fever.

B. She has a cold.

5

A. It is 3 o'clock.

B. It is cool outside.

6

A. The sunset is pretty.

B. The Sun feels warm.

7

A. Turn up the radio.

B. Turn up the heat.

8

A. The water is boiling.

B. The water is clean.

MeasureWorks™ • Grade 1 Temperature • 1

Use Temperature Words

Planning Your Time
Intro & Demo: 15 min
Activity: 10 min
Sum It Up: 5 min

Objective
Review words related to temperature.

Materials
- Classroom thermometer
- Suitcase with different types of clothes
- Crayons

Grouping
Whole class, then individuals or pairs

Open It Up

Tell students that you didn't know what the outside temperature would be today, so you brought clothes for all different temperatures.

Show students the suitcase and its contents: shorts, a sleeveless shirt, a short-sleeved shirt, lightweight pants, a lightweight jacket, mittens, and a winter jacket.

Ask: What are some words we use to describe various outside temperatures? [Samples: **hot, warm, cool, cold**]

What items in my suitcase would be good to wear if it is really hot? [sleeveless shirt, shorts]

What clothes would be good to wear if it is warm? [short-sleeved shirt, lightweight pants or shorts]

What clothes would be good to put on if the weather turns cool? [lightweight jacket]

What clothes would be best if it is really cold? [mittens, winter jacket]

What clothes would be best for today? [Answers will vary.]

Demonstrate & Discuss

Ask: How could I find out what the outside temperature is today? [Samples: read a thermometer or listen to a weather report]

Show the classroom thermometer. Explain that a **thermometer** can be used to find the temperature of the outside air, inside air, liquids, and other things.

Ask: Is ice cold, cool, warm, or hot? [cold]
What is the different between cold and cool? [Cold is colder than cool.] Between warm and hot? [Hot is hotter than warm.]

Student Activity

Prepare ahead: Each student will need crayons.

Read the directions on the student page aloud to students. Read the list of words at the top of the student page aloud together. Then have students work independently to read the questions and mark the answers.

Informal Assessment

As students work, encourage them to use temperature words to describe familiar items.

Ask: What else would you describe as warm (or cool, hot, or cold)? [Sample: my forehead] /COMPREHENSION/

What temperature word would you use to describe a burning candle (or other incorrect answer choice)? [hot] /COMPREHENSION/

Sum It Up

Say: Today we learned about and used words that describe temperature.

Ask: What are some cold (or hot, warm, or cool) things you have seen so far today? [Samples: My milk at lunch was cold. My hot dog was warm. The slide on the playground was hot.] /OBSERVE, DESCRIBE/

Word Study

Make a class book of temperature words. Have each student print a temperature word (*hot, warm, cool,* or *cold*) at the bottom of a sheet of drawing paper. Then have the student draw a picture to represent the word. For the cover, let a volunteer draw and label a thermometer. To assemble the book, punch holes at the tops of the pages and use yarn to tie the pages together.

Temperature • T-2
MeasureWorks™ • Grade 1

Name _____

Hot, Cold, or in Between?

MeasureWorks

Try This

- Here are some temperature words we have talked about.
    cold
    cool
    warm
    hot
    thermometer

Be sure to look at each picture.

- Read each question. Color the picture that answers the question.

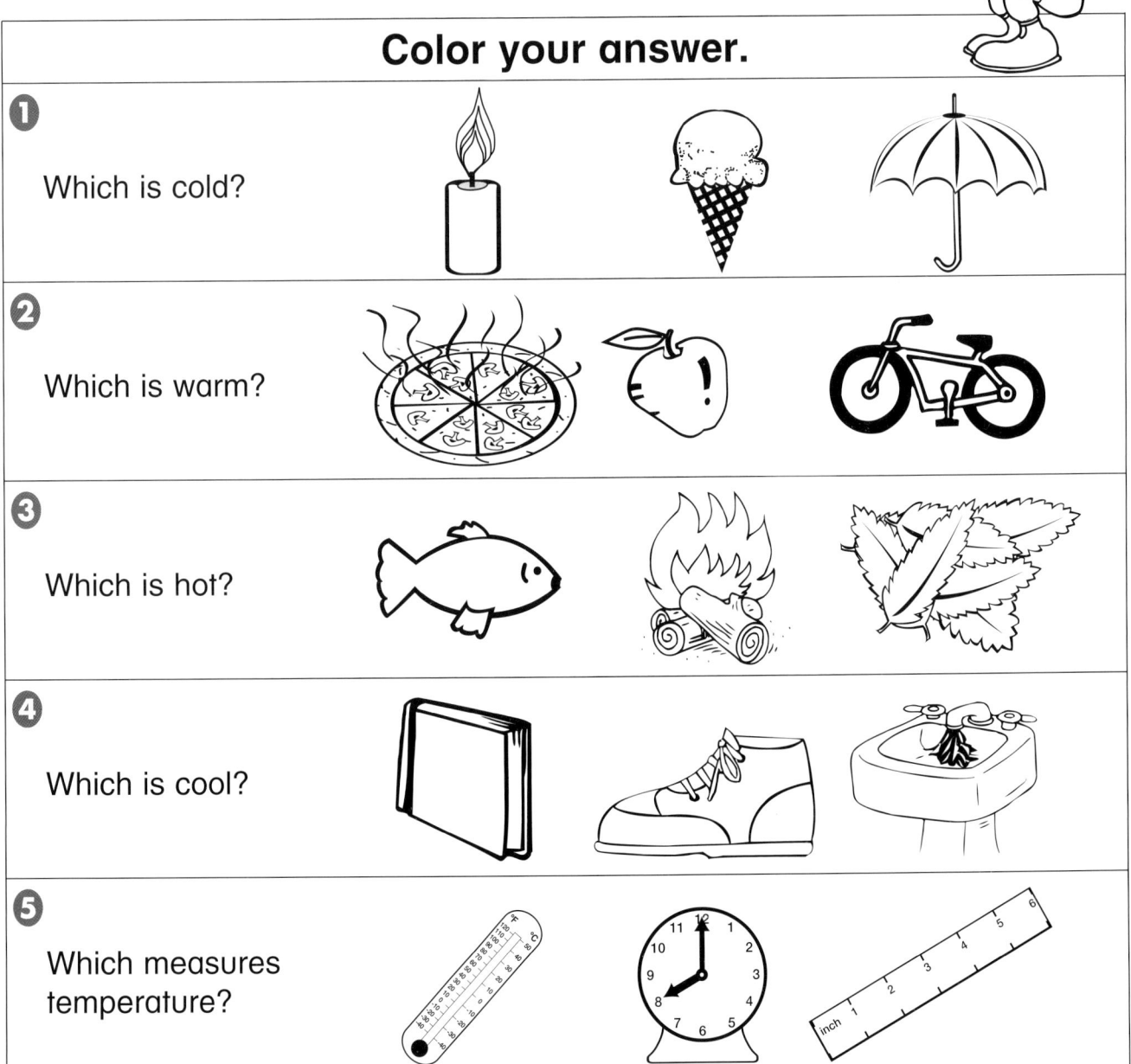

	Color your answer.
1 Which is cold?	candle — ice cream cone — umbrella
2 Which is warm?	pizza — apple — bicycle
3 Which is hot?	fish — fire/log — leaves
4 Which is cool?	book — shoe — sink
5 Which measures temperature?	thermometer — clock — ruler

MeasureWorks™ • Grade 1

Temperature • 2

Compare Temperature

Planning Your Time

Intro & Demo	Activity	Sum It Up
15 min	25 min	5 min

Objective

Compare temperatures.
Order items according to temperature.

Materials

- Paper cups
- Crayons
- Pitcher of water
- Refrigerator or cooler with ice packs

Grouping

Whole class, then pairs

Open It Up

Ask: What does *predict* mean? [It means to tell what you think will happen.]

Tell students that today they will be doing an experiment. When scientists do experiments, they often predict what will happen. Then they do the experiment to see what really happens.

Demonstrate & Discuss

Demonstrate how to set up the experiment. Draw a picture of the Sun on the side of one paper cup. Draw a picture of the refrigerator on the side of a second cup, and draw a picture of a table on a third cup.

Print your initials on the sides of the cups to identify your cups.

Have a volunteer steady the cups as you fill them $\frac{1}{2}$ full with water from a pitcher.

Show students where to set the cups. The cup labeled with a Sun should go on a sunny windowsill. The cup labeled with a refrigerator should go in a refrigerator (or in a cooler with ice packs). The cup labeled with a table should go on a table out of the sun.

Explain that students will predict what the water temperatures will be in a few hours, from **low** (cool) to **high** (warm). Then they will check their predictions.

Student Activity

Prepare ahead: Each pair will need three cups, a pitcher of water, and crayons.

Students work in pairs. Assist the pairs as needed to set up the experiment as you demonstrated. Read the directions on the student page aloud to students. Each student draws pictures across Row 1 of the student page to predict the temperatures, from low to high.

After a few hours, students retrieve their cups, feel the water to compare the temperatures, and arrange the cups from low to high temperature. This step should be done quickly before the water temperatures change.

Each student draws the pictures from the sides of the cups in Row 2 of the student page to show the order from low to high temperature.

Informal Assessment

As students work, encourage them to use temperature terms.

Ask: Why do you think the temperature of the water in the Sun will be higher than the temperature of the water on the table? [The Sun is hot, and it will warm up the water.] /INFER/

What tool could you use to prove that your order is correct? [a thermometer] /COMPREHENSION/

Sum It Up

Say: Today we arranged items according to temperature.

Ask: What were the results in your experiment? [refrigerator, table, Sun] /COMPREHENSION/

Various factors may affect the results of the experiment. If any group has different results, have students suggest possible reasons for the difference.

Science Connection

Let students conduct other experiments involving temperature. For example, if a black container of water and a white container of water are set in the Sun, will the water temperatures differ and if so, which will have the warmer or higher temperature? Will the water temperatures differ if the containers are set in the shade? [Results will vary.]

Name _____

Where Is It Hot?

Measure Works™

Try This

- Draw pictures on the sides of three cups to show where you will put them:

- Fill the cups half-full with water. Put one in the sun, one in a refrigerator, and one on a table.

- Draw pictures to show the temperatures you predict for each cup of water.

Remember: Low temperatures are cold and high temperatures are hot.

❶ What water temperatures do you predict? Show the order from low temperature to high temperature.	**Coolest → in Between → Warmest**

❷ Wait for an hour. Feel the cups again. Draw pictures to show what happened. What happened? Show the order from low temperature to high temperature.	**Coolest → in Between → Warmest**

MeasureWorks™ • Grade 1

Temperature • 3

Order Temperatures

Planning Your Time
Intro & Demo	Activity	Sum It Up
15 min	10 min	5 min

Objective
Compare and order temperatures.
Use temperature vocabulary.

Materials
- Flashlight
- Hair dryer
- Clothes iron

Grouping
Whole class, then individuals

Open It Up

Show students a flashlight. Turn it on and let students feel near the bulb to determine the temperature.

Ask: Is it warm? [yes]

Next, show a hair dryer. Turn on the hair dryer and let students feel the warmth.

Ask: Which of these two things is **warmer**? [the hair dryer]

Finally, show students a clothes iron.
Note: Do not plug in the iron. Tell students that if were turned on, they could not touch it because they might get burned.

Ask: Which of these three things can get **warmest**? [iron]

Demonstrate & Discuss

Ask: What words have we just used to describe and compare temperatures? [warm, warmer, warmest]

List the three words on the board.

Ask: What other words do we use when we compare temperatures? [**hot, hotter, hottest, cold, colder, coldest, cool, cooler, coolest**]

List students' responses on the board.

Student Activity

Read the directions on the student page aloud to students. Guide students as they complete Column 1. Suggest that they begin by locating and labeling the picture that shows the coldest item.

After students have labeled all three of the pictures in Column 1, let them work independently to finish the rest of the page.

Informal Assessment

As students work, encourage them to use the temperature vocabulary words to talk about their work.

Ask: Can you tell me about the temperature words and the pictures in this group? [Sample: The apple is cool. The milk is cooler than the apple. The ice cream is the coolest of all.] /DESCRIBE/

Sum It Up

Say: Today we have used many different words to compare temperatures.

Ask: Are some of the things in group 1 about the same temperature as some of the things in group 2? [Sample: Yes, the ice cream and the pond are both very cold.] /COMPARE AND CONTRAST/

Have volunteers make sentences to compare temperatures of different items, using the words you listed on the board.

Book Nook

On the Same Day in March: A Tour of the World's Weather by Marilyn Singer (HarperCollins, 2000)

This book describes 17 locations around the world on the same day in March. As you read, help the students compare temperatures, from the freezing cold Arctic region to warm, sunny Barbados.

Name _____

Coldest to Hottest

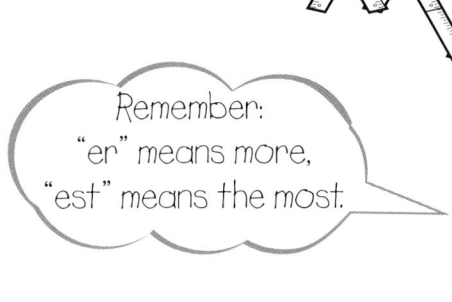

Try This

- Read the temperature words for the first group of pictures.
- Look at the first group of pictures.
- Compare the temperatures.
- Write one of the temperature words under each picture.
- Repeat with each group of pictures.

Remember: "er" means more, "est" means the most.

MeasureWorks

① Cold, Colder, Coldest	② Cool, Cooler, Coolest	③ Warm, Warmer, Warmest	④ Hot, Hotter, Hottest

MeasureWorks™ • Grade 1   Temperature • 4

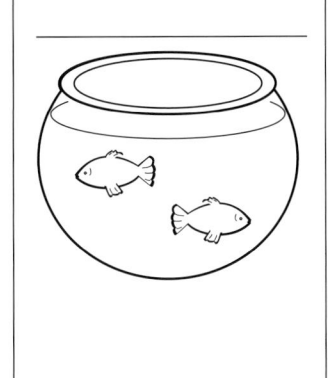

Read a Thermometer

Planning Your Time

Intro & Demo	Activity	Sum It Up
20 min	10 min	5 min

Objective

Discover how a thermometer works.
Read a thermometer (Fahrenheit scale).

Materials

- Classroom thermometer
- Sliding thermometers
- Tub of cold water
- Towel
- Red crayons

Grouping

Whole class, then pairs

Open It Up

Show students the classroom thermometer.

Ask: What is this tool and what does it do? [It is a thermometer, and it measures temperature.]

Explain that the red liquid in the thermometer moves up or down, depending upon the temperature. Put the thermometer into a tub of cold water. Explain that only the bulb (the bottom of the red column) needs to be in the water. Let students watch the red column go down.

Demonstrate & Discuss

Ask: What unit of measure do we use to measure temperature? [**degrees**]

For now, the class will be using the Fahrenheit (F) scale. Explain that the *F* stands for degrees **Fahrenheit,** which is a way of measuring temperature. Explain that the numbers along the red column tell the number of degrees to the nearest ten degrees. Have a volunteer read the number of degrees to the nearest ten. Remove the thermometer from the water and dry it with a towel.

Point out the tiny lines between the numbers on the scale. Explain that each line represents two degrees. Have a volunteer give the exact number of degrees by naming the number printed on the thermometer and then skip counting by two degrees for each additional space.

Point out the negative numbers at the bottom of the thermometers pictured on the page. Explain that the numbers below the zero on a thermometer show very cold temperatures. To read these temperatures, we say, for example, "minus ten degrees" or "ten degrees below zero."

Student Activity

Prepare ahead: Each pair will need a sliding thermometer and two red crayons.

Read the directions on the student page aloud to students.

Students work in pairs. Partners take turns showing the specified temperatures on a sliding thermometer. They they color the thermometers on the student page to show the temperature in degrees.

Informal Assessment

As students work, help them examine and understand details on the thermometers.

Ask: How are the thermometers pictured on the student page different from the classroom thermometer? [Sample: They are smaller.] /COMPARE AND CONTRAST/

Why did you show 64 degrees by coloring two spaces above 60 degrees? [because one mark stands for two degrees] /DESCRIBE/

Sum It Up

Say: Today we learned how to use a thermometer to measure the temperature.

Ask: If the red column of the thermometer goes to the second line above 40 degrees, what is the temperature? [44 degrees] /HYPOTHESIZE/

What is the coldest (hottest) temperature shown on your thermometer? [⁻40 degrees] /OBSERVE/

Extension

Have students work in pairs. Tell the class a temperature. Have each pair show that temperature on the sliding thermometers and hold it up. Have them tell you if the temperature is hot or cold. Play until students are comfortable with the Fahrenheit scale.

Temperature • T-5

MeasureWorks™ • Grade 1

Name _____

Draw Me

Try This

- Read the temperature above the first thermometer in Column 1.
- Show the temperature to your partner on a 🌡.
- Use a red crayon. Color the thermometer to show the correct number of degrees.
- Repeat with the other temperatures and thermometers.

Each line on these thermometers shows 2 degrees.

20°F	−20°F	64°F	98°F

MeasureWorks™ • Grade 1

Temperature • 5

Estimate Temperature

Planning Your Time

Intro & Demo	Activity	Sum It Up
10 min	20 min	5 min

Objective
Estimate water temperatures.
Model and write temperatures.

Materials
- Classroom thermometer
- Sliding thermometers
- Mugs
- Hot tap water
- Ice water
- Lukewarm water
- Refrigerated water

Grouping
Whole class, then small groups

Open It Up

List these three temperatures on the board: 32°F, 70°F, and 212°F.

Ask: Which do you think is the temperature at which water freezes? [32 degrees] Write *water freezes* on the board beside *32°F*.

Which do you think is the temperature at which water boils? [212 degrees] Write *water boils* on the board beside *212°F*.

Which do you think is room temperature? [70 degrees] Write *room temperature* on the board beside *70°F*.

Demonstrate & Discuss

Tell students that an **estimate** is a reasonable guess.

Ask: What would you estimate the temperature of the water in a goldfish bowl to be? [Answers will vary.]

After students have responded, remind them that room temperature is about 70 degrees. Since the goldfish bowl sits in the room day after day, the water in the bowl is likely to be close to the air temperature of the room.

Ask: What would you estimate the temperature of a cup of steaming hot coffee to be and why? [Sample: It would be near 212 degrees because it is made with boiling water, but slightly less because it is no longer boiling.]

Temperature • T-6

Student Activity

Prepare ahead: Each small group will need a sliding thermometer. For the class activity, you will need a mug of hot water, a mug of ice water, a mug of lukewarm water, a mug of refrigerated water, and the classroom thermometer.

Read the directions on the student page aloud to students. Students work in small groups. They estimate the temperature of a cup of hot water. They show their estimates on sliding thermometers. Place a mug of hot tap water on the table. Insert the classroom thermometer. Wait two minutes, and then let groups come forward to read the thermometer. They record the temperature on their student pages. Repeat with the other mugs of water.

Teaching Tip: If other thermometers are available, use them to expedite this activity.

Encourage students to think about the temperatures they have already measured when they make their estimates of temperature.

Informal Assessment

Encourage students to use what they know about temperature to make their estimates.

Ask: Will ice water be colder than 32 degrees? [no] /INFER/

Will lukewarm water be warmer or colder than ice water? [warmer] /COMPARE AND CONTRAST/

Sum It Up

Say: Today we estimated and measured the temperature of water.

Ask: How can knowing the three temperatures listed on the board (32 degrees, 70 degrees, 212 degrees) help you estimate? [Sample: I can compare what I'm measuring to those temperatures.] /GENERALIZE/

Bulletin Board Idea

Make three large paper thermometers for a bulletin board. (Copy and enlarge one of the thermometers on Temperature • T-5.) Let volunteers color the thermometers to show the freezing point of water, the boiling point of water, and room temperature.

MeasureWorks™ • Grade 1

Name _____

About How Hot?

Try This

- Estimate the temperature of hot tap water.
- Show your estimate on a 🌡.
- Measure the temperature to check your estimate.
- Repeat for the other items.

When you estimate, think about the temperatures you have already measured.

Temperature of:	Estimate.	Measure.
❶ Hot tap water	_____°F	_____°F
❷ Ice water	_____°F	_____°F
❸ Lukewarm water	_____°F	_____°F
❹ Refrigerated water	_____°F	_____°F

Did your estimates improve? _____

MeasureWorks™ • Grade 1   Temperature • 6

Measure in Degrees Fahrenheit

Planning Your Time

Intro & Demo	Activity	Sum It Up
15 min	20 min	5 min

Objective
Practice reading a thermometer.
Connect seasons and activities to temperature.

Materials
- Sliding thermometers
- Classroom thermometer
- Pictures showing the four seasons
- Newspaper weather pages for the last few days

Grouping
Whole class, then pairs

Open It Up

Hold up a picture of a summer scene.

Ask: What season does this picture show? [summer] How do you know? [Sample: Students are playing outside in shorts. Flowers are in bloom.] What is the temperature like in summer? [hot] What are some good outside activities for summer? [Samples: swim, play in a sprinkler, ride a bike, swing, play softball.]

Continue by displaying pictures of the remaining seasons and discussing the temperature during each season.

Demonstrate & Discuss

Ask: What is the season now, and what is the temperature usually like in this season? [Answers will vary.]

Show students the newspaper weather pages from the last few days. Locate the high temperature for each date and circle it.

Explain that students will be using a thermometer to find today's temperature. Discuss the fact that the temperature may vary, depending upon the exact location. For example, it may be warmer in the Sun than in the shade.

Teaching Tip: This activity works best on a sunny day. If the weather is too cold to work outdoors, have students measure the temperature in different locations inside (classroom, hall, gym, near a heater, near a window, in a refrigerator, and so on).

Student Activity

Prepare ahead: For this class activity, you will need the classroom thermometer. Each pair will need a sliding thermometer.

Read the directions on the student page aloud to students. As a class, list three outdoor places that might have different temperatures. Possibilities include: in the Sun, in the shade, on the blacktop, in a grassy area, in the wind, or in a protected spot. Then take students outside to measure the temperature. Put the classroom thermometer in each place in turn and allow several minutes for it to adjust to the location. Students read the temperature and record it in the table. In pairs, they take turns showing the temperature on their sliding thermometers.

Teaching Tip: While students wait, you can play "Thermometer Challenge" by announcing a temperature and having them show it on sliding thermometers. Alternatively, you may want to arrange an activity (group game, play on the equipment, relays) to occupy students during the wait time.

Informal Assessment

As students work, encourage them to discuss the use of a thermometer and their results.

Ask: What is the warmest (coolest) place we have found? [Sample: a place in the shade] /COMPARE AND CONTRAST/

Sum It Up

Say: Today we measured the temperature in three different locations.

Ask: Why do you think the temperature was cooler in the shade? [Sample: because shady areas get less sunlight] /HYPOTHESIZE/

Were your temperatures similar to the ones we found in the weather pages? [Sample: No, they were all cooler, because the air is cool this morning.] /COMPARE AND CONTRAST/

Literature Connection

The Seasons of Arnold's Apple Tree by Gail Gibbons (1988, New York: Voyager Books)

This story of an apple tree teaches about the seasons. Read it aloud to students. Then look at the illustrations as you discuss temperature changes throughout the seasons.

Name _____

Explore Outdoors

Temperatures in the shade are usually cool. Temperatures in the Sun are usually warm.

Try This

- List three places outside that might have different temperatures.
- Measure the temperature in each place.
- Record the temperatures.
- Show each temperature on your 🌡.

Place	Temperature
❶	_____ °F
❷	_____ °F
❸	_____ °F

MeasureWorks™ • Grade 1   Temperature • 7

A Week of Weather

Planning Your Time
Wrap Up: 10 min
Project: 60 min

Objective

Measure and record the temperature for one week.
Compare temperatures.

Materials

- KWL charts
- Long range weather forecast for the week
- Classroom thermometer
- Sliding thermometers
- Red colored pencils
- Crayons

Grouping

Whole class, then small groups

Discuss and Sum It Up

* Have students review their KWL charts.
- As a class, discuss what students knew about temperature and what they wanted to know.
- Discuss what students have learned.
- Have students complete their charts.

Assessment

See test on page Assessment • 5.

Project

Prepare ahead: The class will need a long-range weather forecast for the week and the classroom thermometer. Each small group will need sliding thermometers and colored pencils or crayons.

Read the long range weather forecast for the week to the students. Tell students that they will track the weather to see how accurate the forecast proves to be.

Hang the classroom thermometer outside where students can read it. Have students read the temperature each day from Monday to Friday.

Students record the temperature and other information about the day's weather on student page Temperature • 8. Direct students to select and color symbols to show if the day is cloudy, partly cloudy, sunny, windy, rainy, or snowy.

At the end of the week, discuss the week's weather and compare it to the forecast for the week. Have students use red colored pencils to color the thermometers on student page Temperature • 9 to show the week's highest and lowest temperatures.

Finally, ask each student to review the weather for the week and select a favorite day. Each student should use student page Temperature • 10 to record the day and the temperature on that day. They should color the thermometer and draw a picture to show the day's weather. Suggest that students include themselves in their drawings. The picture should show appropriate clothing for the day's weather and an appropriate outside activity.

KWL

I Know	I Want to Know	I Learned
It is hot today.	How hot is it?	One day, it was about 80 degrees Fahrenheit. It was hotter in the Sun than in the shade.
The weather reporter tells the temperature.	How does the weather reporter know the temperature?	People read a thermometer to find the temperature.
Different things are different temperatures.	What temperature are some things in our school?	Air in our classroom is 70 degrees. Cold water from the faucet is about 60 degrees.

Temperature • T-8

MeasureWorks™ • Grade 1

Name _____

A Week of Weather

Weather Record

Each day, read the thermometer.

Record the temperature.

Color the pictures that show the day's weather.

WEATHER REPORT FOR THE WEEK					
Day					
Temperature					
Color the picture that best matches the weather.					

MeasureWorks™ • Grade 1   Temperature • 8

A Week of Weather, continued

High and Low Temperatures

Write the high and low temperatures for the week and on what day.

Show the temperatures on a ⬚.

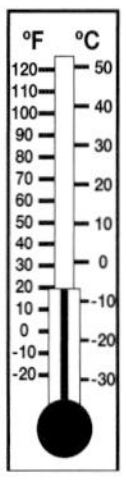

Color the thermometers to show the temperatures.

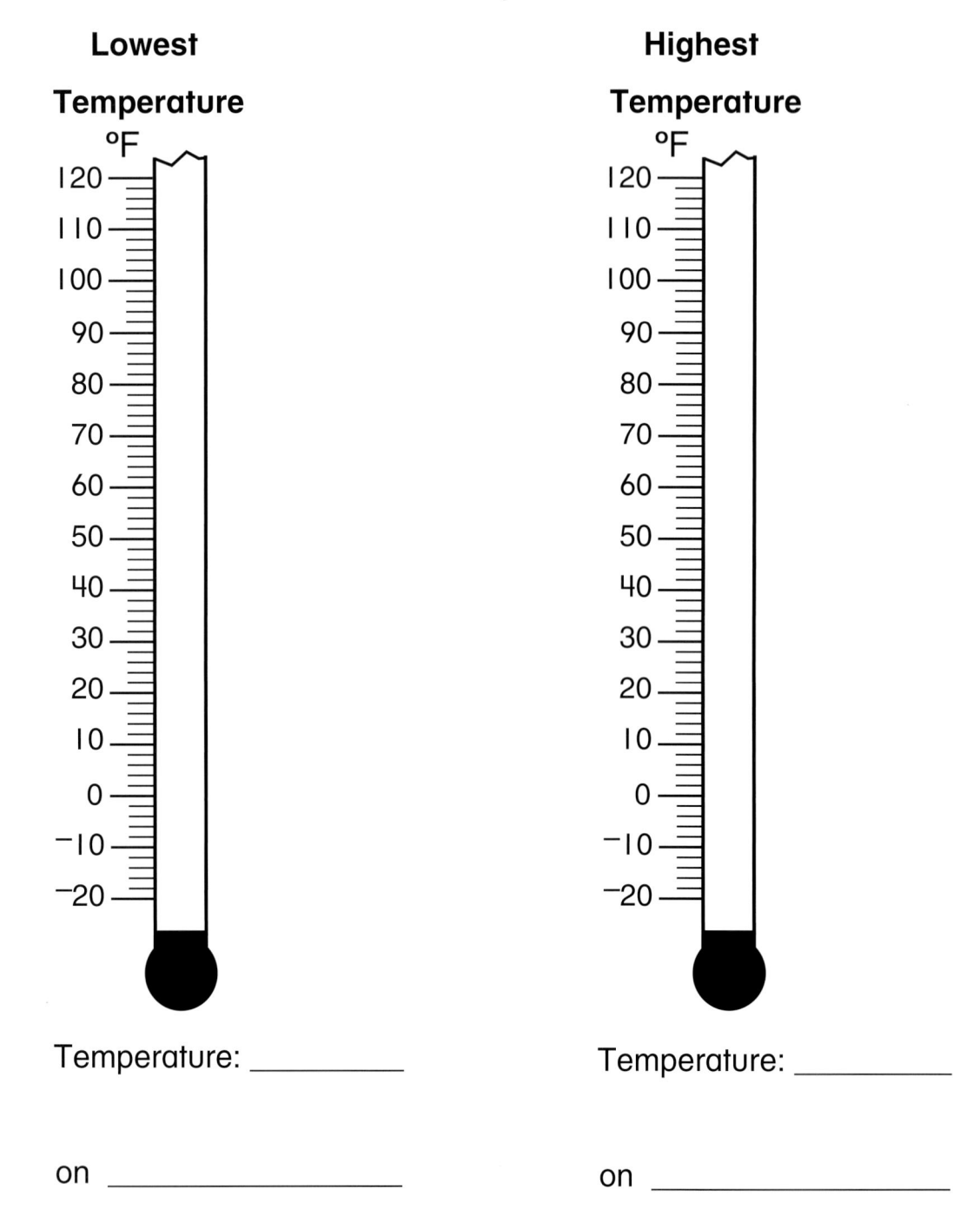

Lowest Temperature
°F

Highest Temperature
°F

Temperature: _____

Temperature: _____

on _____

on _____

Temperature • 9

MeasureWorks™ • Grade 1

A Week of Weather, continued

Favorite Weather

My favorite weather this week was on _____.
(day of week)

1. Write the temperature.
2. Color the thermometer.
3. Draw a picture to show the weather.

MY FAVORITE DAY

Day	
Temperature	

Color the Thermometer

°F
120
110
100
90
80
70
60
50
40
30
20
10
0
−10
−20

Make your own drawing to show what type of day it was.

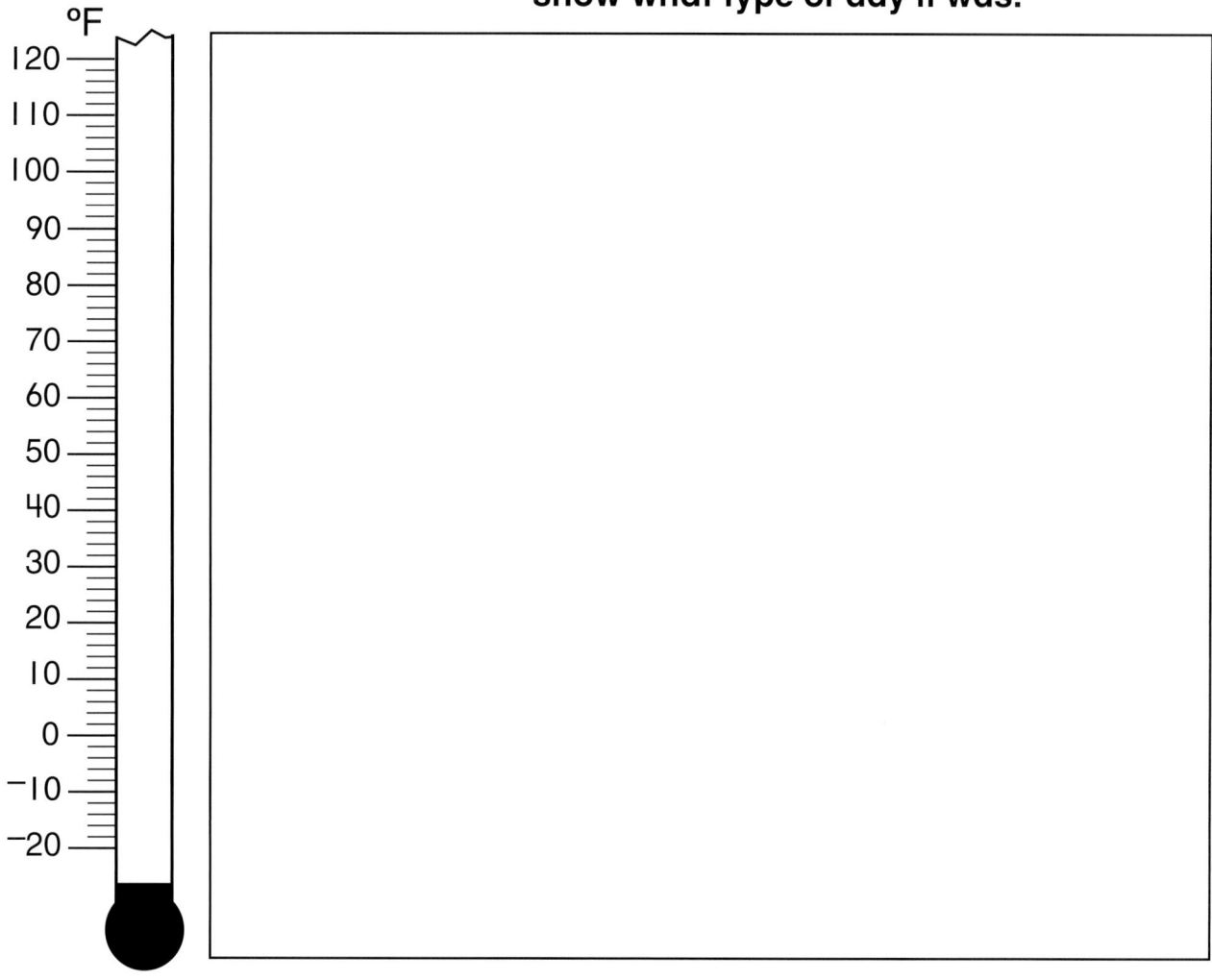

MeasureWorks™ • Grade 1

Answer Key for Temperature Unit

Temperature • 1
1. B
2. B.
3. B
4. A
5. B
6. B
7. B
8. A

Temperature • 2
1. ice cream cone
2. pizza
3. fire
4. water from a faucet
5. thermometer

Temperature • 4
1. coldest, colder, cold
2. cooler, cool, coolest
3. warmest, warmer, warm
4. hot, hottest, hotter

MeasureWorks
Assessment

Time Test

Read each problem aloud. Then read each answer choice together.
Say: Answer each question. Fill in the bubble by your answer choice.
If time allows, have students discuss why the other answer choices are wrong.

Planning Your Time
Intro & SetUp — 5 min
Assess — 15 min
Correct — 20 min

Answer Key and Item Analysis

1
- (A) Student confused the day after with the day before.
- (B) Student did not know the order of the days of the week.
- (C) This choice is correct.
- (D) Student did not know the order of the days of the week.

2
- (F) Student confused the month before February with a month after it.
- (G) Student did not know the order of the months of the year.
- (H) Student did not understand that the question calls for a month before April.
- (J) This choice is correct.

3
- (A) Student may have miscounted the weeks.
- (B) This choice is correct.
- (C) Student chose the wrong day.
- (D) Student may have miscounted the weeks.

4
- (F) This choice is correct.
- (G) Student confused the hour hand and the minute hand.
- (H) Student confused 2 and 3 on the analog clock.
- (J) Student misread the position of the minute hand.

5
- (A) Student confused the position of the hour hand and the minute hand.
- (B) Student misread the position of the hour hand.
- (C) Student did not know how to read the position of the minute hand.
- (D) This choice is correct.

Assessment • T-1

MeasureWorks™ • Grade 1

Name _____

Time Test

Mark the letter of the correct answer.

1 If today is Tuesday, what day is tomorrow?
- (A) Monday
- (B) Thursday
- (C) Wednesday
- (D) Saturday

2 What month is between February and April?
- (F) January
- (G) August
- (H) May
- (J) March

3 What date is the second Monday?
- (A) April 21
- (B) April 14
- (C) April 13
- (D) April 7

APRIL

Sunday	Monday	Tuesday	Wednesday	Thursday	Friday	Saturday
		1	2	3	4	5
6	7	8	9	10	11	12
13	14	15	16	17	18	19
20	21	22	23	24	25	26
27	28	29	30	31		

4 Which digital clock shows the same time as the clock to the right?
- (F) 2:00
- (G) 12:00
- (H) 3:00
- (J) 2:30

5 What time does the clock show?
- (A) 6:00
- (B) 3:30
- (C) 4:15
- (D) 4:30

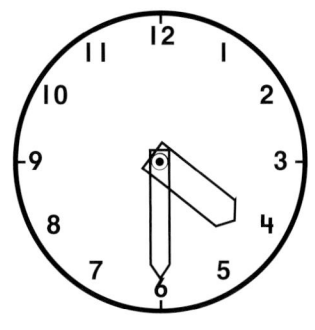

MeasureWorks™ • Grade 1   Assessment • 1

Length Test

Planning Your Time
Intro & SetUp	Assess	Correct
5 min	15 min	20 min

Read each problem aloud. Then read each answer choice together.
Say: Answer each question. Fill in the bubble by your answer choice.
If time allows, have students discuss why the other answer choices are wrong.

Answer Key and Item Analysis

1
- (A) Student did not know how to compare length.
- (B) Student confused longest with shortest.
- (C) This choice is correct.
- (D) Student did not know how to compare length.

2
- (F) Student did not know how to measure with a ruler.
- (G) Student did not pay attention to the units on the ruler.
- (H) Student rounded down instead of up.
- (J) This choice is correct.

3
- (A) Student did not know how to measure with a ruler.
- (B) Student confused the length of the ruler with the length of the straw.
- (C) Student did not know how to measure with a ruler.
- (D) This choice is correct.

4
- (F) This choice is correct.
- (G) Student did not estimate correctly.
- (H) Student confused the number of buttons shown with the estimate.
- (J) Student did not estimate correctly.

Name _____

Length Test

Mark the letter of the correct answer.

1 Which is longest?

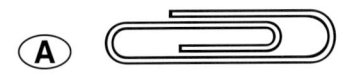

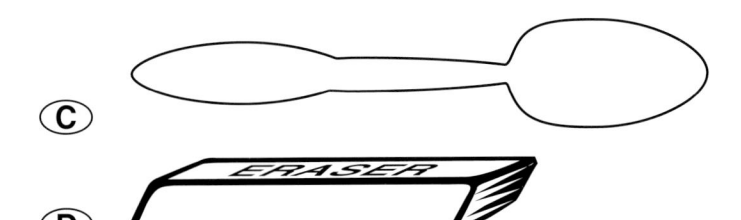

- Ⓐ (paper clip)
- Ⓑ (nail)
- Ⓒ (spoon)
- Ⓓ (eraser)

2 How long is the paper clip? Measure to the nearest inch.

- Ⓕ about 3 inches
- Ⓖ about 2 centimeters
- Ⓗ about 1 inch
- Ⓙ about 2 inches

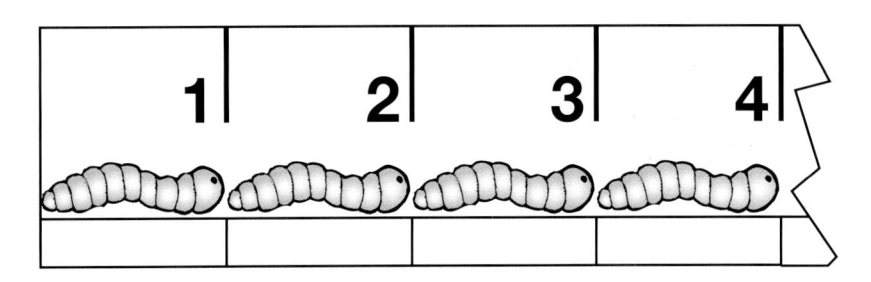

3 How many centimeters long is the straw?

- Ⓐ 5 centimeters
- Ⓑ 8 centimeters
- Ⓒ 7 centimeters
- Ⓓ 6 centimeters

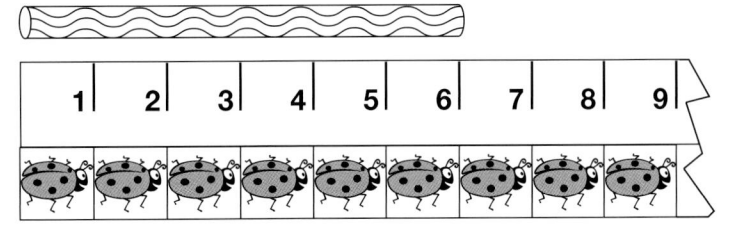

4 Estimate the length of this crayon.

- Ⓕ about 6 buttons long
- Ⓖ about 2 buttons long
- Ⓗ about 1 button long
- Ⓙ about 12 buttons long

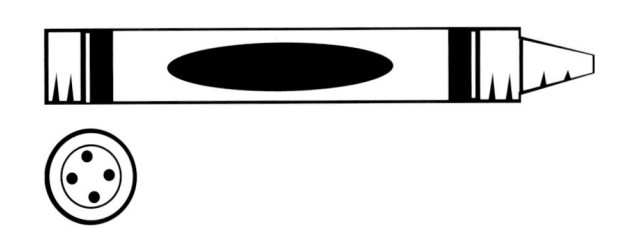

MeasureWorks™ • Grade 1

Assessment • 2

Capacity Test

Planning Your Time
Intro & SetUp: 5 min
Assess: 15 min
Correct: 20 min

Read each problem aloud. Then read each answer choice together.
Say: Answer each question. Fill in the bubble by your answer choice.
If time allows, have students discuss why the other answer choices are wrong.

Answer Key and Item Analysis

1
- (A) This choice is correct.
- (B) Student miscounted or confused more than and less than.
- (C) Student confused more than and less than.
- (D) Student confused more than and less than, or chose a shorter container.

2
- (F) Student may have confused least and most.
- (G) Student may have chosen the tallest container, confusing the size or shape of the container with its capacity.
- (H) This choice is correct.
- (J) Student may have confused the size or shape of the container with its capacity.

3
- (A) This choice is correct.
- (B) Student may have confused the size or shape of the container with its capacity and chosen the thinnest container.
- (C) Student may have confused most and least.
- (D) Student may have chosen the container that appears to be the shortest.

Assessment • T-3

MeasureWorks™ • Grade 1

Name _____

Capacity Test

Mark the letter of the correct answer.

Look at these full containers to answer questions 1–3.

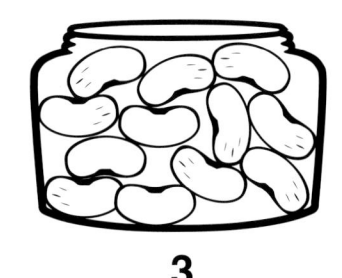

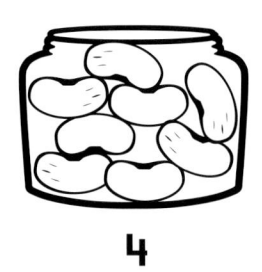

1 **2** **3** **4**

① Which container above holds less than this container?

- Ⓐ Container 1
- Ⓑ Container 2
- Ⓒ Container 3
- Ⓓ Container 4

② Which container holds the most?

- Ⓕ Container 1
- Ⓖ Container 2
- Ⓗ Container 3
- Ⓙ Container 4

③ Which container holds the least?

- Ⓐ Container 1
- Ⓑ Container 2
- Ⓒ Container 3
- Ⓓ Container 4

MeasureWorks™ • Grade 1

Assessment • 3

Weight Test

Read each problem aloud. Then read each answer choice together.
Say: Answer each question. Fill in the bubble by your answer choice.
If time allows, have students discuss why the other answer choices are wrong.

Planning Your Time
Intro & SetUp — 5 min
Assess — 15 min
Correct — 20 min

Answer Key and Item Analysis

1
- Ⓐ Student confused heavier and lighter.
- Ⓑ This choice is correct.
- Ⓒ Student did not understand how to use a balance.
- Ⓓ Student did not understand how to use a balance.

2
- Ⓕ This choice is correct.
- Ⓖ Student confused heavier and lighter.
- Ⓗ Student did not understand how to use a balance.
- Ⓙ Student did not understand how to use a balance.

3
- Ⓐ Student confused the objects to be weighed with their weight.
- Ⓑ This choice is correct.
- Ⓒ Student did not know how to weigh with a balance.
- Ⓓ Student did not look at the unit of measure.

4
- Ⓕ Student did not look at the unit of measure.
- Ⓖ Student did not know how to weigh with a balance.
- Ⓗ This choice is correct.
- Ⓙ Student did not know how to weigh with a balance.

5
- Ⓐ Student did not know how to weigh with a balance.
- Ⓑ This choice is correct.
- Ⓒ Student did not know how to weigh with a balance.
- Ⓓ Student did not know how to weigh with a balance.

Assessment • T-4   MeasureWorks™ • Grade 1

Name _____

Weight Test

Mark the letter of the correct answer.

1 Which object is heavier?

  Ⓐ cup
  Ⓑ glue
  Ⓒ neither
  Ⓓ They weigh the same.

2 Which object is lighter?

  Ⓕ pencil
  Ⓖ milk carton
  Ⓗ neither
  Ⓙ They weigh the same.

3 How much do the apples weigh?

  Ⓐ 2 apples
  Ⓑ 1 pound
  Ⓒ 2 pounds
  Ⓓ 1 kilogram

4 How much does the book weigh?

  Ⓕ 1 pound
  Ⓖ 2 pounds
  Ⓗ 1 kilogram
  Ⓙ 2 kilograms

6 Does the mug weigh more than, less than, or the same as 1 pound?

  Ⓐ more than 1 pound
  Ⓑ less than 1 pound
  Ⓒ the same as 1 pound
  Ⓓ I can't tell.

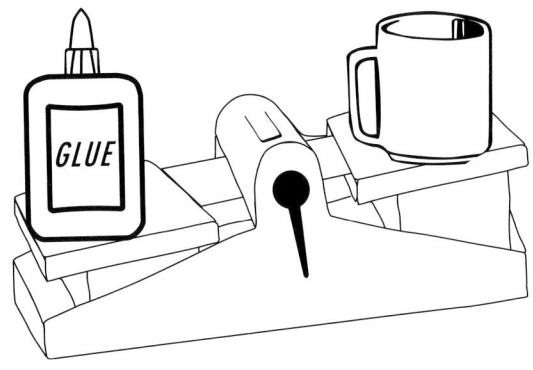

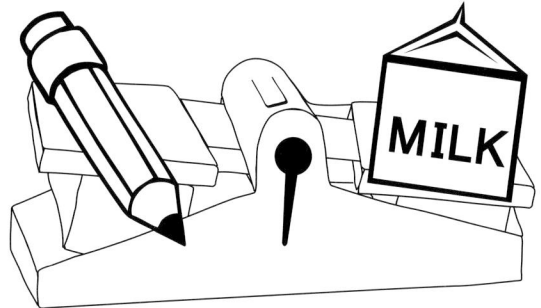

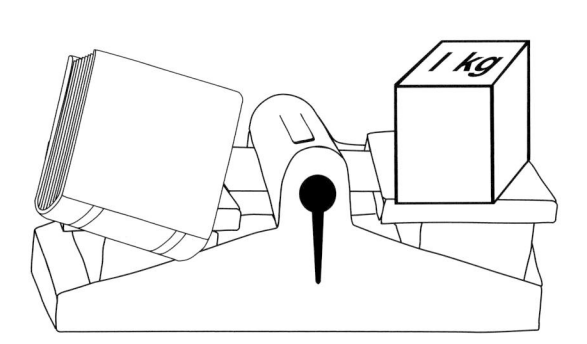

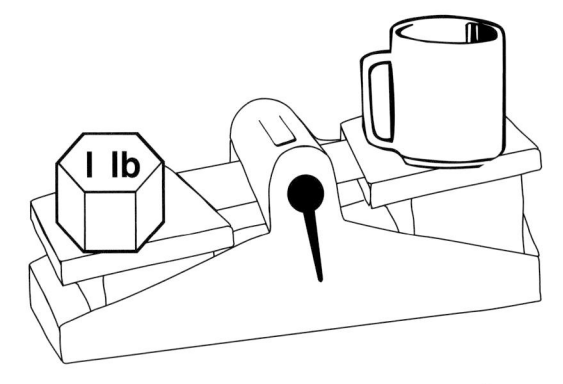

MeasureWorks™ • Grade 1

Assessment • 4

Temperature Test

Read each problem aloud. Then read each answer choice together.
Say: Answer each question. Fill in the bubble by your answer choice.
If time allows, have students discuss why the other answer choices are wrong.

Planning Your Time
Intro & SetUp — 5 min
Assess — 15 min
Correct — 20 min

Answer Key and Item Analysis

1
- (A) Student confused time of day with temperature.
- (B) Student did not recognize that not all weather sentences tell about temperature.
- (C) Student confused a sentence about appearance with temperature.
- **(D) This choice is correct.**

2
- (F) Student did not understand the negative sign.
- (G) Student read the thermometer incorrectly.
- **(H) This choice is correct.**
- (J) Student read the thermometer incorrectly.

3
- **(A) This choice is correct.**
- (B) Student did not know that a higher temperature indicates a warmer temperature.
- (C) Student did not understand the negative sign.
- (D) Student did not know that a higher temperature indicates a warmer temperature.

4
- (F) Student misread the thermometer markings.
- (G) Student misread the thermometer markings.
- (H) Student does not know how to read a thermometer.
- **(J) This choice is correct.**

Assessment • T-5

MeasureWorks™ • Grade 1

Name _____

Temperature Test

Mark the letter of the correct answer.

1 Which sentence is about temperature?

  Ⓐ It is dark outside.

  Ⓑ It is raining outside.

  Ⓒ It is beautiful outside.

  Ⓓ It is warm outside.

2 What temperature does this thermometer show?

  Ⓕ −30 degrees

  Ⓖ 50 degrees

  Ⓗ 30 degrees

  Ⓙ 20 degrees

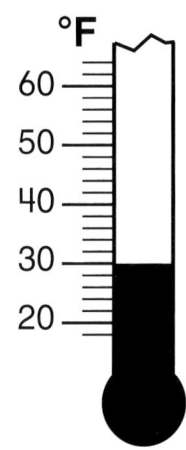

3 Which temperature is the warmest?

  Ⓐ 30°F

  Ⓑ 20°F

  Ⓒ −40°F

  Ⓓ 10°F

4 On Monday, the temperature was about 52°F. Which thermometer shows Monday's temperature?

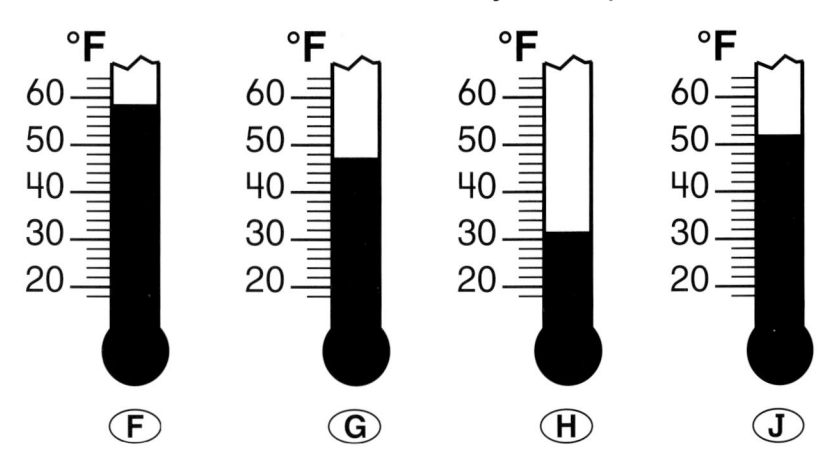

  Ⓕ   Ⓖ   Ⓗ   Ⓙ

MeasureWorks™ • Grade 1   Assessment • 5

Comprehensive Assessment

Read each problem aloud. Then read each answer choice together.
Say: Answer each question. Fill in the bubble by your answer choice.
If time allows, have students discuss why the other answer choices are wrong.

Planning Your Time
Intro & SetUp: 5 min
Assess: 15 min
Correct: 20 min

Answer Key and Item Analysis

1
- (A) Student did not recognize units of time.
- (B) Student may have forgotten about long units of time.
- **(C) This choice is correct.**
- (D) Student did not recognize units of time.

2
- (F) Student chose unit of time that is much too long.
- **(G) This choice is correct.**
- (H) Student chose unit of time that is much too long.
- (J) Student did not recognize a unit of length.

3
- (A) Student confused calendar time and clock time.
- (B) Student did not recognize a tool for measuring temperature.
- (C) Student did not know the function of a schedule.
- **(D) This choice is correct.**

4
- (F) A measuring cup is a tool for measuring liquid, not length.
- **(G) This choice is correct.**
- (H) A balance is a tool for measuring liquid, not length.
- (J) A calendar shows days, weeks, and months. It is not used to measure length.

5
- (A) John would not use hours because he is measuring length, not time.
- (B) John would not use degrees because he is measuring length, not temperature.
- (C) John would not use pounds because he is measuring length, not weight.
- **(D) This choice is correct.**

Assessment • T-6   MeasureWorks™ • Grade 1

Name _____

Comprehensive Assessment

Mark the letter of the correct answer.

1 Which unit is NOT used to measure time?

- Ⓐ minute
- Ⓑ month
- Ⓒ centimeter
- Ⓓ hour

2 How long does it take you to eat lunch?

- Ⓕ 15 days
- Ⓖ 15 minutes
- Ⓗ 15 hours
- Ⓙ 15 meters

3 Dana wants to know if it is after 1:00. Where should she look?

- Ⓐ at a calendar
- Ⓑ at a thermometer
- Ⓒ at a schedule
- Ⓓ at a clock

4 Which tool would be best for measuring the length of this page?

- Ⓕ measuring cup
- Ⓖ ruler
- Ⓗ balance
- Ⓙ calendar

5 John is measuring a marker. How long is it?

- Ⓐ 5 hours
- Ⓑ 32 degrees
- Ⓒ 1 pound
- Ⓓ 12 centimeters

MeasureWorks™ • Grade 1   Assessment • 6

Ⓐ Ⓑ Ⓒ Ⓓ Ⓕ Ⓖ Ⓗ Ⓙ Ⓐ Ⓑ Ⓒ Ⓓ Ⓖ Ⓗ Ⓙ Ⓐ Ⓑ Ⓒ Ⓓ Ⓕ Ⓗ

Comprehensive Assessment

Planning Your Time
Intro & SetUp — 5 min
Assess — 15 min
Correct — 20 min

Read each problem aloud. Then read each answer choice together.
Say: Answer each question. Fill in the bubble by your answer choice.
If time allows, have students discuss why the other answer choices are wrong.

Answer Key and Item Analysis

❶
- Ⓐ This choice is correct.
- Ⓑ Student did not recognize a tool that measures length.
- Ⓒ Student did not recognize a tool that measures time.
- Ⓓ Student did not recognize a tool that measures weight.

❷
- Ⓕ Student did not recognize a unit of length.
- Ⓖ Student did not recognize a unit of weight.
- Ⓗ This choice is correct.
- Ⓙ Student did not recognize a unit of time.

❸
- Ⓐ Student did not recognize a phrase that expresses time.
- Ⓑ This choice is correct.
- Ⓒ Student did not know that weather is not the same as temperature.
- Ⓓ Student did not recognize the temperature of a cold day.

❹
- Ⓕ Student confused units of length with units of weight.
- Ⓖ Student confused units of capacity with units of weight.
- Ⓗ This choice is correct.
- Ⓙ Student confused units of length with units of weight.

❺
- Ⓐ Student confused units of time with units of weight.
- Ⓑ This choice is correct.
- Ⓒ Student confused units of length with units of weight.
- Ⓓ Student confused units of capacity with units of weight.

Assessment • T-7

MeasureWorks™ • Grade 1

Name _____

Comprehensive Assessment
Mark the letter of the correct answer.

1. What tool is used to measure the temperature of the air?
   - Ⓐ thermometer
   - Ⓑ ruler
   - Ⓒ clock
   - Ⓓ balance

2. Jan is reading a thermometer. Which of these might be the temperature?
   - Ⓕ 75 centimeters
   - Ⓖ 75 pounds
   - Ⓗ 75 degrees
   - Ⓙ 75 minutes

3. It is a hot day. What could the temperature be?
   - Ⓐ 12 o'clock noon
   - Ⓑ 80°F
   - Ⓒ 30 inches of rain
   - Ⓓ 20°F

4. Greg weighed his cat. Which unit of measure can Greg use?
   - Ⓕ inches
   - Ⓖ cups
   - Ⓗ kilograms
   - Ⓙ miles

5. Which unit can you use to measure how heavy something is?
   - Ⓐ hours
   - Ⓑ pounds
   - Ⓒ inches
   - Ⓓ cups

MeasureWorks™ • Grade 1   Assessment • 7

Notes

Name: _____   **Review Time Vocabulary**

These words tell about calendar time.

MeasureWorks™ • Grade 1   BLM • 1

Name: _____   **Review Time Vocabulary**

BLM • 2

These words tell about clock time.

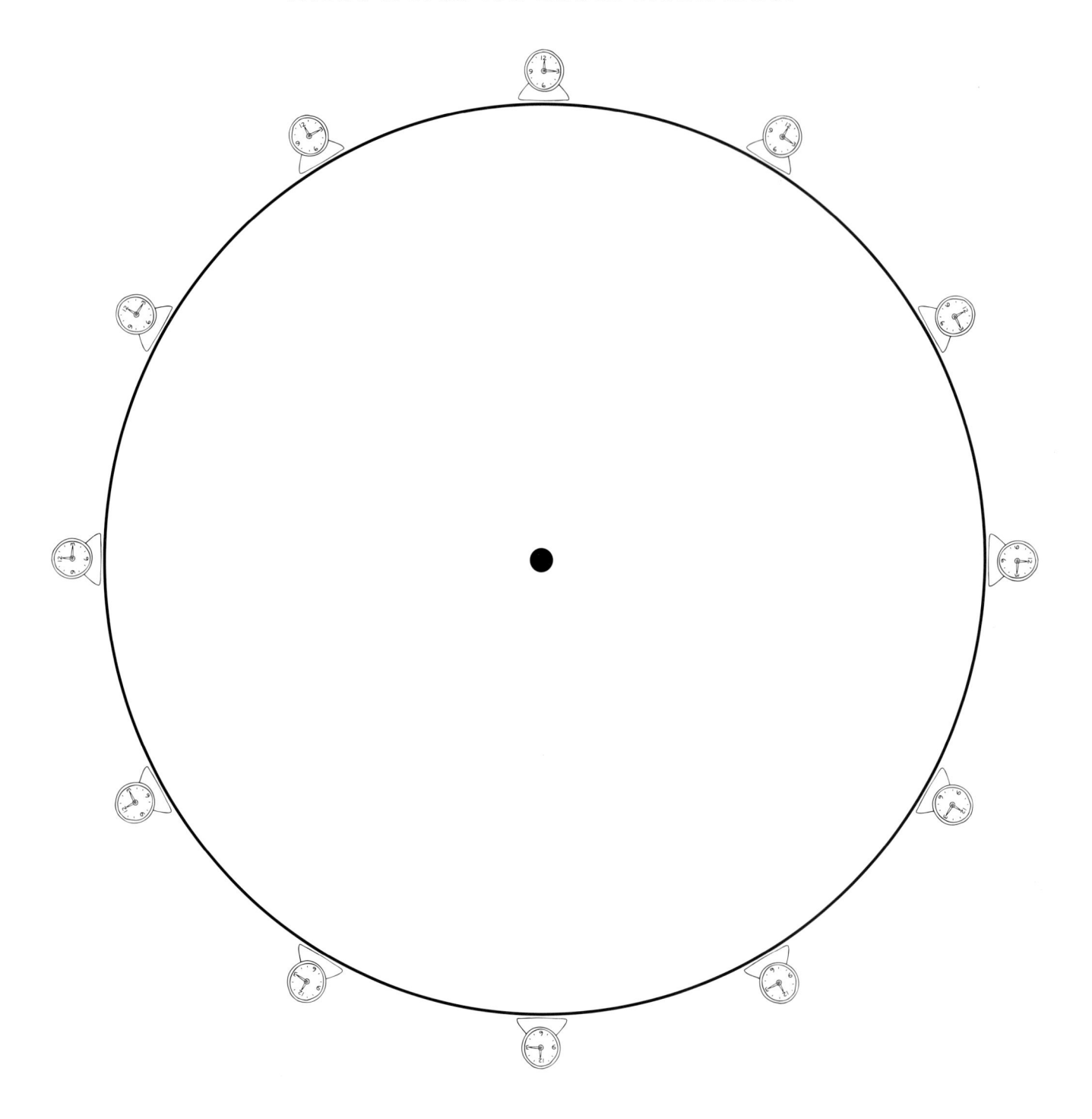

MeasureWorks™ • Grade 1

Name: _____   Days-of-the-Week Memory

MONDAY	TUESDAY	WEDNESDAY
MONDAY	TUESDAY	WEDNESDAY
THURSDAY	FRIDAY	SATURDAY
THURSDAY	FRIDAY	SATURDAY
SUNDAY	TODAY	TOMORROW
SUNDAY	TODAY	TOMORROW
YESTERDAY	THE DAY AFTER TOMORROW	THE DAY BEFORE YESTERDAY
YESTERDAY	THE DAY AFTER TOMORROW	THE DAY BEFORE YESTERDAY

Measurement Master 3

MeasureWorks™ • Grade 1   BLM • 3

Name: _____   **Daily Calendar Activity**

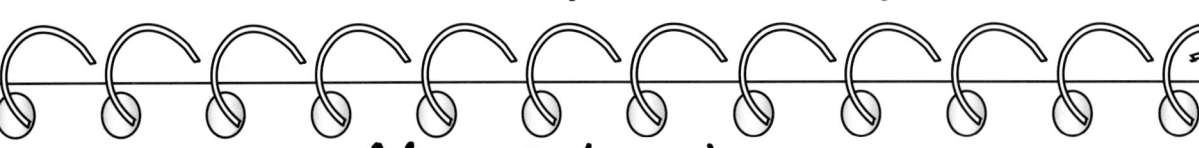

Measurement Master 4

My Calendar

Sunday	Monday	Tuesday	Wednesday	Thursday	Friday	Saturday

MeasureWorks™ • Grade 1

Name: _____　　　**Estimate Time to the Hour**

Measurement Master 5

My Clock

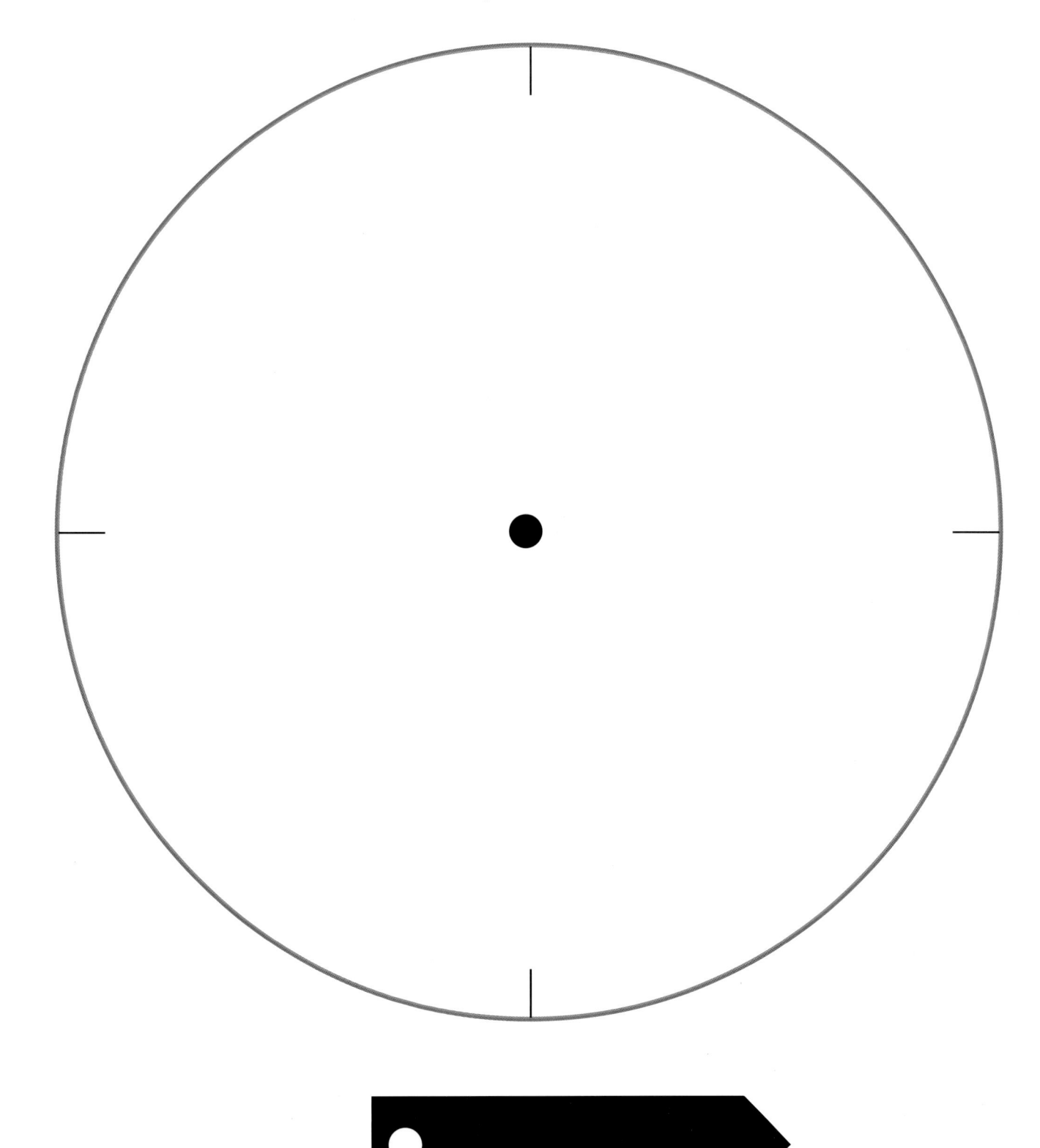

MeasureWorks™ • Grade 1　　　　　　　　　　　　　　　　　　　BLM • 5

Name: _____   **Recognize Time to the Half-Hour**

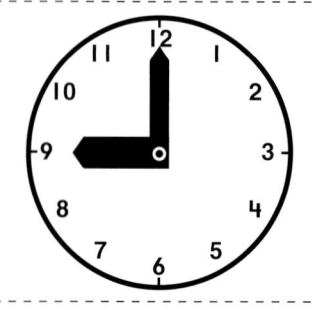

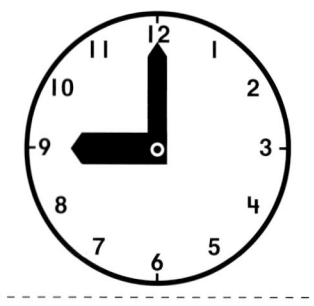

Measurement Master 6

BLM • 6   MeasureWorks™ • Grade 1

Name: _____

KWL

I Know	I Want to Know	I Learned

How to Use a Geared Clock

Clocks help you tell time.

The short red hand tells the hours.

The long blue hand tells the minutes.

When the long blue hand points straight up, the time is zero minutes past the hour. We say "o'clock" to tell time to the hour.

Now look more closely.

Each blue dot marks one minute.

The blue numbers skip count minutes by 5.

The minute hand tells how many minutes past the hour.

This clock shows 15 minutes past 8.

BLM • 8

MeasureWorks™ • Grade 1

How to Measure to the Nearest Inch

Get to know your ruler.

One side shows Inchworms. Each worm is one <u>inch</u> long.

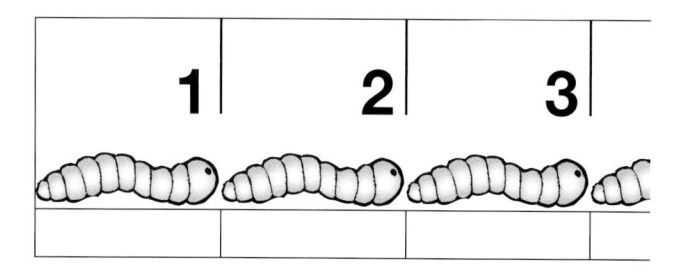

To measure length, line up one end of your object with the left edge of the ruler. Look at the ruler where the object ends.

This fat pencil is about 3 inches long.

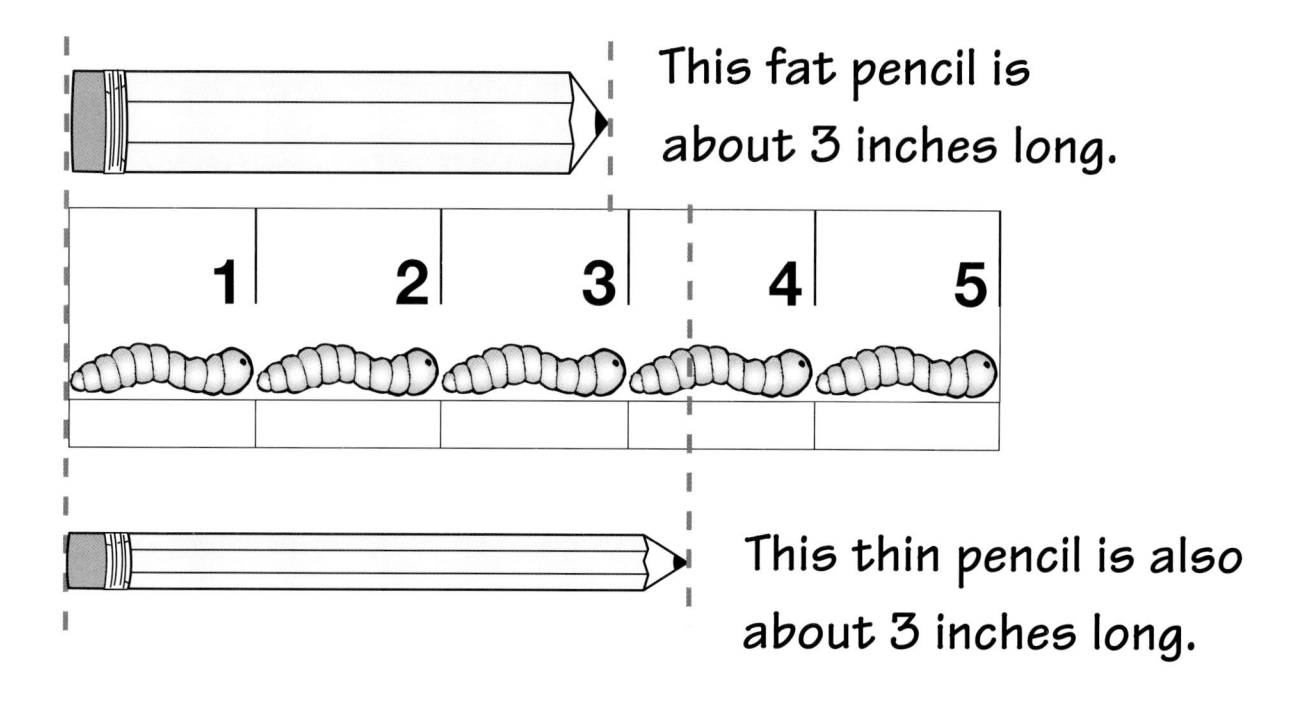

This thin pencil is also about 3 inches long.

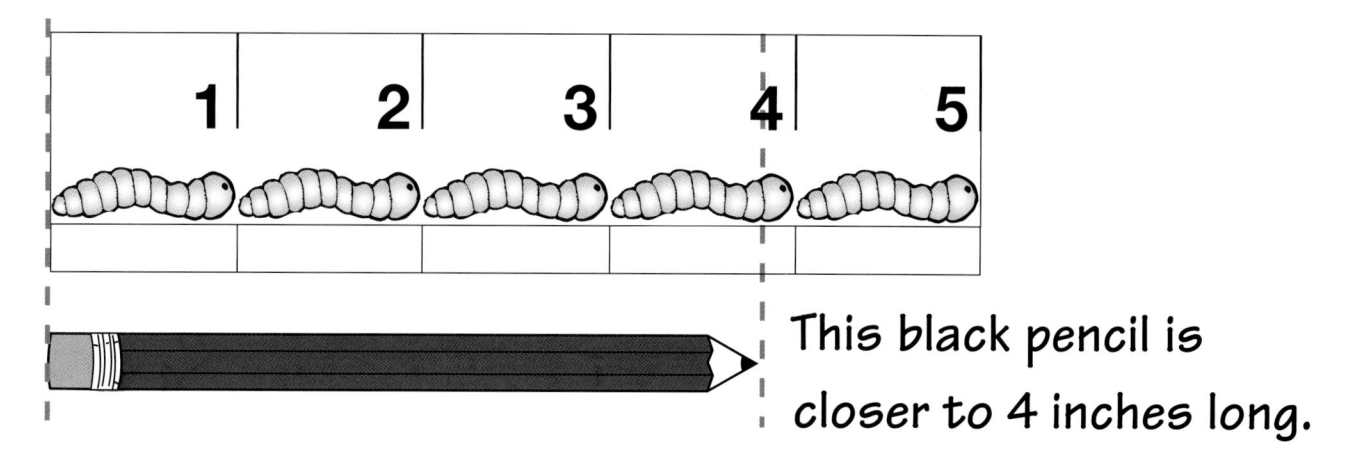

This black pencil is closer to 4 inches long.

MeasureWorks™ • Grade 1

How to Measure to the Nearest Centimeter

Look at your ruler.
Find the side that shows centibugs.

Each bug is one <u>centimeter</u> long.

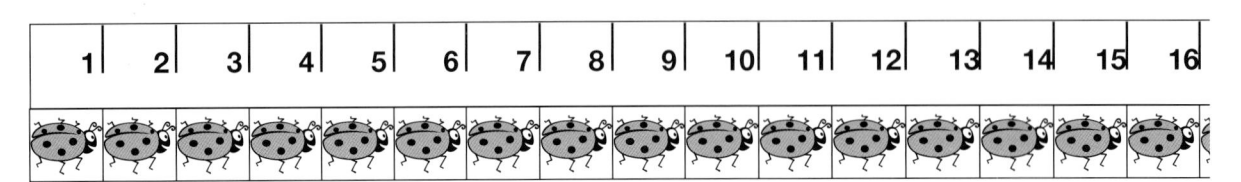

You can measure anything to the nearest centimeter.

Follow these steps.

Step 1:
Line up edges.

Step 2:
Find the end.

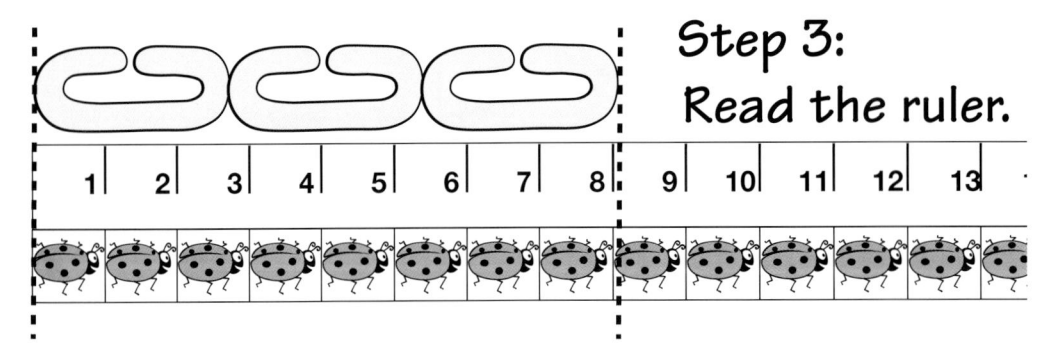

Step 3:
Read the ruler.

Three links together are about 8 centimeters long.

How to Use a Pan Balance

A pan balance compares the weight of two objects.

When the pointer points straight down, the objects weigh the same.

Set the balance on a level surface.

Check that the pointer points straight down. Adjust if needed.

Put an object in the center of one pan.

Notice how the pointer changes position.

Put an object in the center of the other pan.

Notice how the pointer changes position.

The pan balance shows that the apple is heavier than the bear.

How a Thermometer Works

A thermometer measures temperature in degrees.

Heat makes the liquid in the tube rise.

Room temperature is about 68°F.

Today is a hot day! It is 30°C. That's about 87°F.

CLASSROOM THERMOMETER

°F   °C
120 — 50
110 —
100 — 40
90 —
80 — 30
70 — 20
60 —
50 — 10
40 —
30 — 0
20 —
10 — -10
0 —
-10 — -20
-20 — -30
-30 —
-40 — -40

°F means degrees Fahrenheit.

°C means degrees Celsius.

Cold makes the liquid in the tube fall.

Room temperature is about 20°C.

Water freezes at 0°C. That's 32°F.

BLM • 12   MeasureWorks™ • Grade 1

MeasureWorks Game Rules

Object of Game A • Train Station

Be the first to answer a question correctly in each measurement strand and answer a game-winning question at Home.

About Game Board A

Game board A is made up of nine sections with a path that connects the sections to each other. There are five strand sites that players must visit during each game. Each site represents a different measurement strand, as shown below.

- **Volumetric Labs** (Volume and Capacity)
- **Home Dimensions Hardware** (Length)
- **Sub-Weigh Deli** (Weight)
- **Cool 'N' Hot Sports** (Temperature)
- **Train Station** (Time)

Note: Grade 1 does not use the Perimeter Patch.

The color of each space on the board represents a different measurement strand, as shown below.

- **Pink**—Volume/Capacity
- **Orange**—Length
- **Green**—Weight
- **White**—Temperature
- **Light Green**—Time
- **Yellow, Blue**—Player's Choice

MeasureWorks™ • Grade 1   BLM • 13

Short Cuts for Fame A

The two sections of the board that show the railroad tracks can be used to take a shortcut across the board. A player who lands in either section may automatically move to the opposite section that shows railroad tracks during the same turn.

There are two staircases shown on the board. A player who lands on either space that shows a staircase may automatically move to the opposite staircase during the same turn.

Prepare to Play Train Station

Stack the MeasureWorks cards facedown in a pile on the table. Choose a People Counter and place it in the Home area of the game board. Have each player roll the die. The player who rolls the highest number goes first.

On Your Turn

❶ Roll the die and move your People Counter out of the Home area that number of spaces. You may move your People Counter in either direction on the path.

Note: You are not allowed to go backward over spaces you have already crossed on that turn. For example, if your People Counter is three spaces from a strand site and you roll a 5, you cannot move four spaces forward and one space back to land on the entrance to the site.

❷ When you land on a space, another player draws the top card from the deck and reads you the question for that color. When you have answered, put the card at the bottom of the deck. If you land on a yellow square or a blue square, choose any category you like.

Note: The next question will be read from the next card in the deck, and so on. When all the cards have been used, shuffle the cards and place them facedown in a stack again.

❸ If you answer the question correctly, your People Counter remains on that space. If you answer the question incorrectly, move your People Counter back to its previous position. After you answer your question, play passes to the player on your left.

❹ When you land on a strand site entrance, move your People Counter into the site. If you answer the question correctly, you earn a PopCube of that color. If you answer incorrectly, your turn ends. On your next turn, you may roll and move, or you may stay in the same strand site and answer another question in that strand to try to win the PopCube.

❺ If you land on Home before you have earned all five PopCubes, you may choose the category your question will come from—so pick one you're good at!

On Your Turn, continued

6 If you pass through Home during your move, count it as one space.

7 Any number of People Counters may occupy the same space at the same time.

8 After you have correctly answered a question in all five strand sites, head for Home. You do not have to roll the exact count to land on Home if you have all five PopCubes—a number larger than what you need is OK. For example, if you roll a 5 and need only a 3, you can stop in Home and forget about the extra 2. Remember that in all other cases, you have to move the exact number of spaces shown on the die.

Winning Game A

- When you've earned all five PopCubes and have reached Home to try to answer the game-winning question, the other players choose the strand for your question.

- If you answer incorrectly, you can stay in Home and answer a question on your next turn—until you get one right and win.

- The winner is the first player to answer a question correctly in all five categories and answer the game-winning question at Home.

Notes

- The rules do not state how much time you have to answer a question or how exact the answer must be. The players decide this themselves.

- Guessing is better than not answering at all. You probably know a lot more than you think you do, so take a guess!

MeasureWorks Game A • Train Station

Collect five 🧊 and answer a question at Home.

- Start at Home. Take turns.

- Roll the die. Move your 🧍.

- When you land, answer a question for that color.
- If you land on yellow, blue, or Home, choose a color.

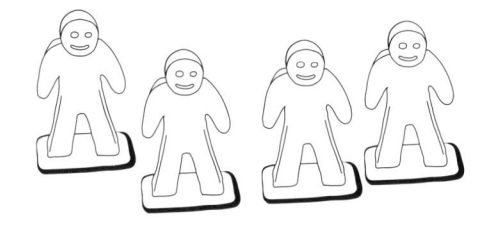

If you know the answer, stay. If not, move back.

- When you land on "Enter Here," go in the space.

If you know the answer, take a 🧊. If not, try again on your next turn.

- Get one 🧊 of each color.

- When you have five 🧊, move to Home.

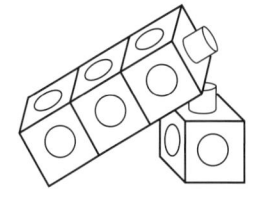

⭐ **To Win:** When you get to Home, the other players read a question. If you know the answer, you win. If not, try again on your next turn.

MeasureWorks™ • Grade 1 　　　　　　　　　　　　　　　　BLM • 17

Object of Game B • Theme Park Game

Be the first to travel completely around the game board and land back at the roller coaster depot.

About Game Board B

Game board B is a roller coaster ride that consists of a depot (Home) and a series of roller-coaster cars that represent the spaces on the board.
The color of each space on the board represents a different measurement strand, as shown below.

- **Blue**—Area/Perimeter
- **Pink**—Volume/Capacity
- **Orange**—Length
- **Green**—Weight
- **White**—Temperature
- **Light Green**—Time
- **Yellow**—Angles (Grades 4 and 5 only)

Note: Grades 1–3 players may answer questions from any measurement strand when they land on a yellow space.

Note: Grade 1 players answer questions from any measurement topic when they land on a blue space.

Prepare to Play Theme Park

Stack the MeasureWorks cards facedown in a pile on the table. Choose a People Counter and place it in the roller coaster depot (Home) on the game board. Have each player roll the die. The player who rolls the highest number goes first.

On Your Turn

1 Roll the die and move your People Counter that number of spaces.

2 When you land on a space, another player draws the top card from the deck and reads you the question for that color. When you have answered, put the card at the bottom of the deck. The next question will be read from the next card in the deck, and so on. When all the cards have been used, shuffle the cards and place them face down in a stack again.

3 If you answer your question correctly, your People Counter remains on that space. If you answer incorrectly, move your People Counter back to its previous position. After you answer your question, play passes to the player on your left.

4 Any number of People Counters may occupy the same space at the same time.

5 You do not have to roll the exact count to land on the depot (Home) on the game board. For example, if you roll a 5 and need only a 3, you can stop on the depot and forget about the extra 2.

Tips

The rules do not state how much time you have to answer a question or how exact the answer must be. The players decide this themselves.

Guessing is better than not answering at all. You probably know a lot more than you think you do, so take a guess!

Notes

Notes